# The Transformation of Intergovernmental Satellite Organisations

# Studies in Space Law



VOLUME 9

*The titles published in this series are listed at brill.com/slaw*

# The Transformation of Intergovernmental Satellite Organisations

## Policy and Legal Perspectives

*Edited by*

Patricia K. McCormick
Maury J. Mechanick

MARTINUS
NIJHOFF
PUBLISHERS

LEIDEN · BOSTON
2013

Library of Congress Cataloging-in-Publication Data

The transformation of intergovernmental satellite organisations : policy and legal perspectives / edited by Patricia K. McCormick, Maury J. Mechanick.
pages cm. -- (Studies in space law ; volume 9)
Includes bibliographical references and index.
ISBN 978-90-04-25706-1 (hardback : alk. paper) 1. Artificial satellites in telecommunication--Law and legislation. I. McCormick, Patricia K., editor of compilation. II. Mechanick, Maury J., editor of compilation.

K4307.T73 2013
384.5'1--dc23

2013027347

This publication has been typeset in the multilingual "Brill" typeface. With over 5,100 characters covering Latin, IPA, Greek, and Cyrillic, this typeface is especially suitable for use in the humanities. For more information, please see www.brill.com/brill-typeface.

ISSN 1871-7659
ISBN 978-90-04-25706-1 (hardback)
ISBN 978-90-04-25707-8 (e-book)



This book is printed on acid-free paper.

# Contents

# Preface

More than a decade has now passed since the implementation of the decision to privatise or restructure the world's most prominent intergovernmental satellite organisations, namely, the International Telecommunications Satellite Organisation (INTELSAT), the International Mobile Satellite Organisation (INMARSAT) and the European Telecommunication Satellite Organisation (EUTELSAT). The ramifications of these notable alterations of international institutions have yet to be fully evaluated. This book, *The Transformation of Intergovernmental Satellite Organisations: Policy and Legal Perspectives*, is intended to fill this void by analyzing the complicated processes and associated corollaries of the structural reform of these international satellite organisations. It is further considered that the perspectives acquired from the passage of time since these decisions were undertaken facilitate this research endeavour, as distance fosters a clearer assessment of the procedures pursued and their resultant effects.

The first chapter, "Neo-liberalism: A Contextual Framework for Assessing the Privatisation of Intergovernmental Satellite Organisations," seeks to situate the privatisation and restructuring of the ISOs within the larger context of neo-liberalism. In the span of a few short years, these intergovernmental satellite organisations were essentially disbanded as they were simultaneously transformed into private corporate enterprises, owned, no longer by governments and by extension, citizens, but now, in some cases, by mutual fund management companies. This is an expansion of the application of neo-liberal principles, which in the 1990s was evident in the global diffusion of policy to reform and privatise national telecommunications entities. In tracing the evolution of neo-liberal tenets from the 1980s onward, noting its affects on the global economy, developing states, and international institutions, this chapter seeks to employ an international political-economic framework to critically examine the transformation of the former intergovernmental satellite organisations.

The second chapter, "INMARSAT: In the Forefront of Mobile Satellite Communications," examines INMARSAT's transformation into a private national corporate entity as provider of its services, with an intergovernmental organisation, the International Mobile Satellite Organisation (IMSO), acting as a regulator of certain services deemed vital to the international public interest including those relating to the Global Maritime Distress and Safety System (GMDSS). In 1998 INMARSAT's Assembly of member Governments agreed to amend the INMARSAT Convention and Operating Agreement to permit the restructuring and privatisation of INMARSAT, the first intergovernmental organisation to be privatised. As a global intergovernmental organisation (IGO) serving public interests, the transformation of INMARSAT into a private sector corporation, with an international regulator, was fairly unique in international corporate life. The process took some nine years, and in many respects served as a model for INTELSAT and EUTELSAT. Throughout its structural alteration, INMARSAT has served the mariner, the aviator and the land based user with high quality communications, using modern mobile satellite telecommunications technology. This chapter outlines the background and special features of the INMARSAT organisation, the particular forces which led inexorably to its transformation, the key elements of the new structure, the differing interests of its membership and the main political, commercial and legal issues encountered.

The third chapter, "INTELSAT: Pre and Post-Private Equity Ownership," examines the privatisation of INTELSAT, which underwent a profound restructuring in November 2000 when the 143 Member States of INTELSAT made the historic decision to restructure the organisation into two separate entities. The actual satellite system of INTELSAT was privatised by transferring substantially all of its assets and liabilities, including satellites and orbital fillings, to a new company established for this purpose, Intelsat, Ltd., incorporated in Bermuda, and its subsidiaries. The other entity, retaining the name International Telecommunications Satellite Organisation (ITSO), is a highly modified revision of the intergovernmental organisation initially designated to serve for twelve years (with its mandate having recently been extended through at least 2021, representing an additional eight-year period) in a supervisory and monitoring capacity to ensure that the private company meets its public service and lifetime connectivity obligations to those developing countries largely dependent on the services and pricing mechanisms of Intelsat. In tracing the historical development of INTELSAT, this chapter critically analyzes the complicated consequences of the notable structural and ownership alterations of INTELSAT and the influence exerted by Lockheed Martin Corporation in the restructuring process, as well as US legislation, specifically the Open-market Reorganisation for the Betterment of International Telecommunications Act or ORBIT Act, which set forth specific criteria relating to the privatisation of INTELSAT, fundamentally altering its intergovernmental treaty arrangement.

The fourth chapter, "The Evolution of EUTELSAT: A Challenge Successfully Met," presents an examination of the European Telecommunication Satellite Organisation before, during and after its restructuring. The term "restructuring" is used in this analysis, as opposed to that of "privatisation," since privatisation generally refers to the sale of publicly owned interests to a private entity. The restructuring of EUTELSAT consisted of maintaining an intergovernmental organisation, EUTELSAT IGO, albeit with a transformed structure, role, mission and activities focused on ensuring that the principles of pan-European coverage, fair competition, and universal service are met; and transferring all of the assets, operational activities and related liabilities and commitments of an existing intergovernmental organisation (EUTELSAT) to a private entity (Eutelsat S.A.) established in France for this purpose. Between 2002 and 2005, however, all the major telecommunication operators sold their shares to financial investors which then resold them to other financial investors. In 2005, following a leverage recapitalisation and the creation of the holding company Eutelsat Communications S.A., more than 95 percent of the shareholding of Eutelsat S.A. was held by the holding company, which was floated on the Paris Stock exchange. In January and February 2007, two long term investors, namely the Spanish group Abertis and the French entity CDC ("Caisse des dépôts et consignations") became major shareholders, and, to date, this ownership structure remains. This chapter details these and other alterations to the ISO as well as developments in its service provision.

The fifth chapter, "The Role and Function of Residual International Intergovernmental Satellite Organisations Following Privatisation," considers how public policy issues were addressed throughout the privatisation process for each of the ISOs. The privatisation of the three ISOs are structurally similar in that intergovernmental agencies or organisations (IMSO, ITSO, EUTELSAT IGO) were transformed to ensure that the private companies, to which all assets and activities were respectively transferred, adhere to public service standards. This chapter considers the rationale for the preservation of some form of intergovernmental oversight over the privatised satellite companies that emerged from the privatisation process; the scope of authority that was conferred on those oversight bodies; the anticipated duration of their continued existence and mechanisms for their funding; their role and responsibilities – if any – with respect to the orbital resources previously held by the ISOs; the manner in which the public service obligations were to be discharged; and various internal governance issues. It also examines the relationship between these oversight bodies and other international organisations. The chapter concludes with an assessment of the functioning and success of these oversight bodies to date.

The sixth chapter, "Crossing a Rubycon? The International Legal Framework for ISOs—Before and After Privatisation," focuses on a legal analysis of the activities of international satellite organisations (ISOs), former ISOs and private satellite operators from the perspective of international space law. It provides a comprehensive examination of the application of the Outer Space Treaty, the Rescue and Return Agreement, the Liability Convention and the Registration Convention, as well as the UN Resolution on Broadcasting and – to some extent – the ITU regime.

This book, which draws on some of the best legal minds in the field of international telecommunications and satellite law, provides readers with new depths of insight and historical context for the restructuring of the former intergovernmental satellite organisations. This analysis can be employed to better assess international governance issues in an era of greater private sector participation across all sectors and institutions. Indeed, in this regard, it is hoped that public service is not considered as simply obligatory, but undertaken with a view to truly improve communications within and between states.

I would like to express my sincere appreciation to the contributors whose efforts and persistence created this work. In addition, I would like to extend a special note of gratitude to the Series Editor, Frans von der Dunk, and co-editor, Maury Mechanick, both of whom provided continued support and encouragement throughout the process.

Patricia K. McCormick

# List of Key Legal Documents

## 1. *International Treaties*

1944

*Chicago Convention:* Convention on International Civil Aviation, Chicago, done 7 December 1944, entered into force 4 April 1947; 15 UNTS 295; TIAS 1591; 61 Stat. 1180; Cmd. 6614; UKTS 1953 No. 8; ATS 1957 No. 5; ICAO Doc. 7300.

1945

*UN Charter:* Charter of the United Nations, San Francisco, done 26 June 1945, entered into force 24 October 1945; USTS 993; 24 UST 2225; 59 Stat. 1031; 145 UKTS 805; UKTS 1946 No. 67; Cmd. 6666 & 6711; CTS 1945 No. 7; ATS 1945 No. 1.

1946

Convention on the Privileges and Immunities of the United Nations, London, done 13 February 1946, entered into force 17 September 1946; 1 UNTS 15; UKTS 10 (1950) Cmd. 7891; ATS 1949 No. 3.

1947

Agreement between the United Nations and the United States of America regarding the Headquarters of the United Nations, Lake Success, done 26 June 1947, entered into force 21 November 1947; 11 UNTS 11; 554 UNTS 308 (1966); 687 UNTS 408 (1969).

*GATT:* General Agreement on Tariffs and Trade, Geneva, done 30 October 1947, entered into force 1 January 1948; 55 UNTS 194; TIAS 1700; ATS 1948 No. 23.

## 1948

*IMCO Convention:* Convention on the Intergovernmental Maritime Consultative Organisation IMCO (hereafter IMCO Convention), Geneva, done 6 March 1948, entered into force 17 March 1958; 289 UNTS 48; TIAS 4044; UKTS 1958 No. 54; Cmnd. 589; Cmd. 7412; ATS 1958 No. 5; Title of Convention was amended to "Convention on the International Maritime Organisation" in 1975 with effect from 22 May 1982.

## 1949

*First Geneva Convention:* Geneva Convention for the Amelioration of the Condition of the Wounded and Sick in Armed Forces in the Field, Geneva, done 12 August 1949, entered into force 21 October 1950; 75 UNTS 31; TIAS No. 3362; 6 UST 3114; ATS 1958 No. 21.

*Second Geneva Convention:* Geneva Convention for the Amelioration of the Condition of the Wounded, Sick and Shipwrecked Members of Armed Forces at Sea, Geneva, done 12 August 1949, entered into force 21 October 1950; 75 UNTS 85; TIAS No. 3363; 6 UST 3217; ATS 1958 No. 21.

*Third Geneva Convention:* Geneva Convention relative to the Treatment of Prisoners of War, Geneva, done 12 August 1949, entered into force 21 October 1950; 75 UNTS 135; TIAS No. 3364; 3 UST 3316; ATS 1958 No. 21.

*Fourth Geneva Convention*: Geneva Convention relative to the Protection of Civilian Persons in Time of War, Geneva, done 12 August 1949, entered into force 21 October 1950; 75 UNTS 287; TIAS No. 3365; 3 UST 3516; ATS 1958 No. 21.

## 1967

*Outer Space Treaty:* Treaty on Principles Governing the Activities of States in the Exploration and Use of Outer Space, including the Moon and Other Celestial Bodies, London/Moscow/Washington, done 27 January 1967, entered into force 10 October 1967; 610 UNTS 205; TIAS 6347; 18 UST 2410; UKTS 1968 No. 10; Cmnd. 3198; ATS 1967 No. 24; 6 ILM 386 (1967).

1968

*Rescue Agreement:* Agreement on the Rescue of Astronauts, the Return of Astronauts and the Return of Objects Launched into Outer Space, London/Moscow/Washington, done 22 April 1968, entered into force 3 December 1968; 672 UNTS 119; TIAS 6599; 19 UST 7570; UKTS 1969 No. 56; Cmnd. 3786; ATS 1986 No. 8; 7 ILM 151 (1968).

1969

Vienna Convention on the Law of Treaties, Vienna, done 23 May 1969, entered into force 27 January 1980; 1155 UNTS 331; UKTS 1980 No. 58; Cmnd. 4818; ATS 1974 No. 2; 8 ILM 679 (1969).

1971

*INTELSAT Agreement:* Agreement Relating to the International Telecommunications Satellite Organisation (INTELSAT), Washington, done 20 August 1971, entered into force 12 February 1973; 1220 UNTS 21; TIAS 7532; 23 UST 3813; UKTS 1973 No. 80; Cmnd. 4799; ATS 1973 No. 6; 10 ILM 909 (1971).

*INTELSAT Operating Agreement:* Operating Agreement Relating to the International Telecommunications Satellite Organisation (INTELSAT), Washington, done 20 August 1971, entered into force 12 February 1973; 1220 UNTS 149; TIAS 7532; 23 UST 4091; UKTS 1973 No. 80; Cmnd. 4799; ATS 1973 No. 6; 10 ILM 946 (1971).

*INTERSPUTNIK Agreement:* Agreement on the Establishment of the "INTERSPUTNIK" International System and Organisation of Space Communications, Moscow, done 15 November 1971, entered into force 12 July 1972; 862 UNTS 3; *Space Law – Basic Legal Documents*, C.VIII.1.

1972

*Liability Convention:* Convention on International Liability for Damage Caused by Space Objects, London/Moscow/Washington, done 29 March 1972, entered into force 1 September 1972; 961 UNTS 187; TIAS 7762; 24 UST 2389; UKTS 1974 No. 16; Cmnd. 5068; ATS 1975 No. 5; 10 ILM 965 (1971).

1974

*SOLAS Convention:* International Convention for Safety of Life at Sea, London, done 1 November 1974, entered into force 25 May 1980; 1184 UNTS 278, 1300 UNTS 391, 1408 UNTS 339, 1484 UNTS 442 & 1593 UNTS 417; TIAS 9700 & 10626; UKTS 1980 No. 46 & UKTS 1983 No. 42; ATS 1983 No. 22.

1975

*Registration Convention:* Convention on Registration of Objects Launched into Outer Space, New York, done 14 January 1975, entered into force 15 September 1976; 1023 UNTS 15; TIAS 8480; 28 UST 695; UKTS 1978 No. 70; Cmnd. 6256; ATS 1986 No. 5; 14 ILM 43 (1975).

*ESA Convention:* Convention for the Establishment of a European Space Agency, Paris, done 30 May 1975, entered into force 30 October 1980; 14 ILM 864 (1975); *Space Law – Basic Legal Documents*, C.I.1.

1976

Agreement of the Arab Corporation for Space Communications (ARABSAT), Cairo, done 14 April 1976, entered into force 15 July 1976; *Space Law – Basic Legal Documents*, C.VII.1; 44 *Telecommunications Journal* (IX/1977), at 422.

*INMARSAT Convention:* Convention on the International Maritime Satellite Organisation (INMARSAT), London, done 3 September 1976, entered into force 16 July 1979; 1143 UNTS 105; TIAS 9605; 31 UST 1; UKTS 1979 No. 94; Cmnd. 6822; ATS 1979 No. 10; 15 ILM 1052 (1976).

*INMARSAT Operating Agreement:* Operating Agreement on the International Maritime Satellite Organisation (INMARSAT), London, done 3 September 1976, entered into force 16 July 1979; 1143 UNTS 213; TIAS 9605; 31 UST 1; UKTS 1979 No. 94; Cmnd. 6822; ATS 1979 No. 10; 15 ILM 233, 1075 (1976).

1977

Agreement on the Constitution of a Provisional European Telecommunications Satellite Organisation "INTERIM EUTELSAT", Paris, done 13 May 1977, entered into force 30 June 1977; ESA, *Basic Texts*, Vol. III, Doc. K6.

*Geneva Protocol I:* Protocol Additional to the Geneva Conventions of 12 August 1949, and relating to the Protection of Victims of International Armed Conflicts, Geneva, done 8 June 1977, entered into force 7 December 1978; ATS 1991 No. 29; 16 ILM 1391 (1977); 72 *American Journal of International Law* (1978), 457.

*Geneva Protocol II:* Protocol Additional to the Geneva Conventions of 12 August 1949, and relating to the Protection of Victims of Non-International Armed Conflicts, Geneva, done 8 June 1977, entered into force 7 December 1978; ATS 1991 No. 30; 16 ILM 1442 (1977); 72 *American Journal of International Law* (1978), 502.

1978

*ECS Agreement:* Supplementary Agreement Relating to the Space Segment of the Satellite Telecommunications System for the Fixed Service (ECS), Paris, done 10 March 1978, entered into force 14 September 1978; ESA, *Basic Texts*, Vol. III, Doc. K7; *Aerospace Law – Telecommunications Satellites* (1982), 322.

1979

*Moon Agreement:* Agreement Governing the Activities of States on the Moon and Other Celestial Bodies, New York, done 18 December 1979, entered into force 11 July 1984; 1363 UNTS 3; ATS 1986 No. 14; 18 ILM 1434 (1979).

1982

*EUTELSAT Convention:* Convention Establishing the European Telecommunications Satellite Organisation (EUTELSAT), Paris, done 15 July 1982, entered into force 1 September 1985; Cmnd. 9069; *Space Law – Basic Legal Documents*, C.II.1.

*EUTELSAT Operating Agreement:* Operating Agreement Relating to the European Telecommunications Satellite Organisation (EUTELSAT), Paris, done 15 July 1982, entered into force 1 September 1985; Cmnd. 9154; *Space Law – Basic Legal Documents*, C.II.2.

1986

Vienna Convention on the law of treaties between states and international organisations or between international organisations, Vienna, done 21 March 1986, not yet entered into force; Cm. 244; 25 ILM 543 (1986).

1992

*EC Treaty:* Treaty establishing the European Economic Community, Rome, done 25 March 1957, entered into force 1 January 1958; 298 UNTS 11; as amended by the Treaty on European Union, Maastricht, done 7 February 1992, entered into force 1 November 1993; 31 ILM 247 (1992); OJ C 191/1 (1992).

*EEA Agreement:* Agreement on the European Economic Area, Oporto, done 2 May 1992, entered into force 1 January 1994; OJ L 1/3 (1994).

*ITU Constitution:* Constitution of the International Telecommunication Union, Geneva, done 22 December 1992, entered into force 1 July 1994; 1825 UNTS 1; UKTS 1996 No. 24; Cm. 2539; ATS 1994 No. 28; Final Acts of the Additional Plenipotentiary Conference, Geneva, 1992 (1993), at 1.

*ITU Convention:* Convention of the International Telecommunication Union, Geneva, done 22 December 1992, entered into force 1 July 1994; 1825 UNTS 1; UKTS 1996 No. 24; Cm. 2539; ATS 1994 No. 28; Final Acts of the Additional Plenipotentiary Conference, Geneva, 1992 (1993), at 71.

1994

*WTO Agreement:* Agreement Establishing the World Trade Organisation, Marrakesh, done 15 April 1994, entered into force 1 January 1995; 1867 UNTS; UKTS 1996 No. 57; ATS 1995 No. 8; 33 ILM 1125, 1144 (1994).

*GATS*: General Agreement on Trade in Services, Marrakesh, done 15 April 1994, entered into force 1 January 1995; ATS 1995 No. 8.

*TRIPS Agreement:* Agreement on Trade-Related Aspects of Intellectual Property Rights, Annex 1C of the GATS, Marrakesh, done 15 April 1994, entered into force 1 January 1995; ATS 1995 No. 38.

Instrument amending the Constitution of the International Telecommunication Union (Geneva, 1992), Kyoto, done 14 October 1994, entered into force 1 January 1996; Cm. 3447; ATS 1996 No. 10; Final Acts of the Plenipotentiary Conference, Kyoto, 1994 (1995), at 1.

Instrument amending the Convention of the International Telecommunication Union (Geneva, 1992), Kyoto, done 14 October 1994, entered into force 1 January 1996; Cm. 3447; ATS 1996 No. 10; Final Acts of the Plenipotentiary Conference, Kyoto, 1994 (1995), at 23.

## 1996

Protocol on the Amendments to the Agreement on the Establishment of the "INTERSPUTNIK" International System and Organisation of Space Communications, done November 1996, entered into force 4 November 2002; *Space Law – Basic Legal Documents*, C.VIII.2.

## 1997

Agreement on Telecommunications Services, Geneva, done 15 February 1997, entered into force 5 February 1998; ATS 1998 No. 9; 36 ILM 354 (1997).

Fourth Protocol to the General Agreement on Trade and Services of 15 April 1994, Geneva, done 15 April 1997, entered into force 5 February 1998; ATS 1998 No. 9; 33 ILM 1167 (1994); 36 ILM 354 (1997).

## 1998

Instrument amending the Constitution of the International Telecommunication Union of 22 December 1992, as amended 14 October 1994, Minneapolis, done, 6 November 1998, entered into force 1 January 2000; ATS 2000 No. 8.

Instrument amending the Convention of the International Telecommunication Union of 22 December 1992, as amended 14 October 1994, Minneapolis, done, 6 November 1998, entered into force 1 January 2000; ATS 2000 No. 8.

*IMSO Convention:* Convention on the International Mobile Satellite Organisation, London, done 3 September 1976, entered into force 16 July 1979, as amended 1998, amended version applied provisionally 15 April 1999, entered into force 31 July 2001; ATS 2001 No. 11.

1999

*Public Services Agreement:* Public Services Agreement Between the International Mobile Satellite Organisation And Inmarsat One Limited And Inmarsat Two Company, London, done April 1999.

*EUTELSAT Convention as amended:* Convention Establishing the European Telecommunications Satellite Organisation (EUTELSAT), Paris, done 15 July 1982, entered into force 1 September 1985, as amended 20 May 1999, amended version applied provisionally 2 July 2001, entered into force 28 November 2002; *Space Law – Basic Legal Documents*, C.II.1.

2000

*ITSO Agreement:* Agreement Relating to the International Telecommunications Organisation (ITSO), Washington, done 20 August 1971, entered into force 12 February 1973, as amended 17 November 2000, amended version applied provisionally 18 July 2001, entered into force 30 November 2004; *Space Law – Basic Legal Documents*, C.V.1.

## 2. *International Jurisprudence*

1928

*Case concerning Chorzów Factory:* Case concerning the factory at Chorzów (Merits)(Germany v. Poland), Permanent Court of International Justice, 13 September 1928, P.C.I.J., Ser. A, No. 17.

1955

*Nottebohm Case:* Nottebohm Case (Second Phase)(Liechtenstein v. Guatemala), International Court of Justice, 6 April 1955, I.C.J. Rep. 1955, 4.

1970

*Barcelona Traction Case:* Case Concerning the Barcelona Traction Light and Power Company, Limited (Second Phase)(Belgium v. Spain), International Court of Justice, 5 February 1970, I.C.J. Rep. 1970, 4.

## 3. *United Nations Resolutions*

1961

UNGA Res. 1721(XVI) B, of 20 December 1961; General Assembly – Sixteenth Session, Resolutions adopted on reports of the First Committee, at 6.

1982

Principles Governing the Use by States of Artificial Earth Satellites for International Direct Television Broadcasting, UNGA Res. 37/92, of 10 December 1982; UN Doc. A/AC.105/572/Rev.1, at 39.

2000

United Nations Millennium Declaration, UNGA Res. 55/2, of 18 September 2000; UN Doc. A/RES/55/2.

2010

Resolution on international cooperation in the peaceful uses of outer space, UNGA Res. 64/86, of 13 January 2010; UN Doc. A/RES/64/86.

## 4. *Documents Other International Organisations*

ESA

Declaration of 23 September 1976 (hereafter ESA Declaration on Liability Convention); *International Organisations and Space Law* (1999), p. 33; *Space Law – Basic Legal Documents*, A.III.2, at 1.

Resolution on the Agency's Legal Liability, ESA/C/XXII/Res. 3, adopted Paris, 13 December 1977; *International Organisations and Space Law* (1999), p. 35.

IMO

Resolution A.888(21), Criteria for the Provision of Mobile-Satellite Communication Systems in the Global Maritime Distress and Safety System (GMDSS), of 25 November 1999; I:ASSEMBLY21Res888.doc.

Resolution MSC 275(85), of 5 December 2008; I:MSC8526-Add-1.doc.

ITU

Declaration of the First Meeting of Equatorial Countries of 3 December 1976, ITU Doc. WARC-BS (1977) 81-E, of 17 January 1977; *Space Law – Basic Legal Documents*, B.IV.1; 6 *Journal of Space Law* (1978), 193.

## 5. *European Community / European Union Legislation*

1962

Regulation (EEC) No. 17/62, of 6 February 1962, OJ 13/204 (1962).

1989

Directive of the European Parliament and of the Council on the coordination of certain provisions laid down by law, regulation or administrative action in Member States concerning the provision of audiovisual media services (Audiovisual Media Services Directive), 89/552/EEC, of 3 October 1989; OJ L 298/23 (1989).

1994

Commission Directive amending Directive 88/301/EEC and Directive 90/388/EEC in particular with regard to satellite communications, 94/46/EC, of 13 October 1994; OJ L 268/15 (1994).

1997

Directive of the European Parliament and of the Council amending Council Directive 89/552/EEC on the coordination of certain provisions laid down by law, regulation or administrative action in Member States concerning the pursuit of television broadcasting activities, 97/36/EC, of 30 June 1997; OJ L 202/60 (1997).

2002

*Radio Spectrum Decision:* Decision of the European Parliament and of the Council on a regulatory framework for radio spectrum policy in the

European Community, No. 676/2002/EC, of 7 March 2002; OJ L 108/1 (2002).

*Access Directive*: Directive of the European Parliament and of the Council on access to, and interconnection of, electronic communications networks and associated facilities, 2002/19/EC, of 7 March 2002; OJ L 108/7 (2002).

*Authorisation Directive*: Directive of the European Parliament and of the Council on the authorisation of electronic communications networks and services, 2002/20/EC, of 7 March 2002; OJ L 108/21 (2002).

*Framework Directive:* Directive of the European Parliament and of the Council on a common regulatory framework for electronic communications networks and services, 2002/21/EC, of 7 March 2002; OJ L 108/33 (2002).

*Universal Service Directive:* Directive of the European Parliament and of the Council on universal service and users' rights relating to electronic communications networks and services, 2002/22/EC, of 7 March 2002; OJ L 108/51 (2002).

Directive of the European Parliament and of the Council concerning the processing of personal data and the protection of privacy in the electronic communications sector, 2002/58/EC, of 12 July 2002; OJ L 201/37 (2002).

2007

Directive of the European Parliament and of the Council amending Council Directive 89/552/EEC on the coordination of certain provisions laid down by law, regulation or administrative action in Member States concerning the pursuit of television broadcasting activities, 2007/65/EC, of 11 December 2007; OJ L 332/27 (2007).

## 6. *Other documents European Community / European Union*

1990

*Green Paper on Satellite Telecommunications:* Towards Europe-wide systems and services – Green Paper on a common approach in the field of satellite communications in the European Community, Communication from the Commission, COM(90) 490 final, of 20 November 1990; ... (1990).

## 7. *National Legislation*

France

Act n°66-537 of 24 July 1966 on commercial companies.
*Loi Toubon*; Act n° 94-665 of 4 August 1994 concerning the use of the French language.
Act n°2004-575 of 21 June 2004 on confidentiality in the digital economy.
Act n°2004-669 of 9 July 2004 concerning electronic communications and audiovisual services.
Decree n°2007-1532 of 24 October 2007 concerning charges for the utilisation of radio-electric frequencies payable by holders of licences issued by the regulatory authority for electronic communications and post.
*Loi relative aux opérations spatiales*; Loi n° 2008-518 du 3 juin 2008; 34 *Journal of Space Law* (2008), at 453; unofficial translation 34 *Journal of Space Law* (2008), at 453.
Decision n° 2008-656 of 8 July 2008 extending the licence issued to SA LV & Co. concerning the utilisation of a category D radio service by terrestrial hertzian means with modulating frequency (MFM).
Act n° 2009-258 of 5 March 2009 concerning audiovisual communication and a new public television service.
Decree n° 2009-640 of 9 June 2009 on the application of the conditions of chapter VII of Act n° 2008-518 of 3 June 2008 concerning space operations.
Decree n° 2009-643 of 9 June 2009 concerning licences issued in accordance with Act n° 2008-518 of 3 June 2008 concerning space operations.

United Kingdom

Telecommunications Act, 12 April 1984, 1984 Chapter 12.
Outer Space Act, 18 July 1986, 1986 Chapter 38; *National Space Legislation of the World*, Vol. I (2001), at 293; *Space Law – Basic Legal Documents*, E.I; 36 *Zeitschrift für Luft- und Weltraumrecht* (1987), at 12.

United States of America

Communications Act, 19 June 1934; 47 U.S.C. 151 (1988); 48 Stat. 1064.
Communications Satellite Act, Public Law 87-624, 87th Congress, H.R. 11040, 31 August 1962; 76 Stat. 419; as amended 1978; *Space Law – Basic Legal Documents*, E.III.2.

Communications Satellite Facilities, *First Report and Order*, 22 FCC 2d 86 (1970), Appendix C, p. 1.

Arms Export Control Act of 1976, 22 U.S.C. 2751.

Export Administration Act of 1979; Public Law 96-72, 96th Congress; 50 U.S.C. 2401; 93 Stat. 503.

Land Remote-Sensing Commercialization Act, Public Law 98-365, 98th Congress, H.R. 5155, 17 July 1984; 98 Stat. 451; *Space Law – Basic Legal Documents*, E.III.4.

Commercial Space Launch Act, Public Law 98-575, 98th Congress, H.R. 3942, 30 October 1984; 98 Stat. 3055; *Space Law – Basic Legal Documents*, E.III.3.

Commercial Space Launch Act Amendments, Public Law 100-657, 100th Congress, H.R. 4399, 15 November 1988; 49 U.S.C. App. 2615; 102 Stat. 3900; *Space Law – Basic Legal Documents*, E.III.3, 13 ff.

Land Remote Sensing Policy Act, Public Law 102-555, 102nd Congress, H.R. 6133, 28 October 1992; 15 U.S.C. 5601; 106 Stat. 4163.

Commercial Space Transportation – Commercial Space Launch Activities, 49 U.S.C. 70101 (1994).

Telecommunications Act, Public Law 104-104, 104th Congress, 3 January 1996, signed into law 8 February 1996; 110 Stat. 56.

Strom Thurmond National Defense Authorisation Act for Fiscal Year 1999; Public Law 105-261, 105th Congress, 17 October 1998; 112 Stat. 1920.

Federal Communications Commission, *In the Matter of Direct Access to the INTELSAT System. IB Docket No. 98-192. File No. 60-SAT-ISP-97. Report and Order.* Adopted 15 September 1999.

*ORBIT Act:* Open-market Reorganisation for the Betterment of International Telecommunications Act, Public Law 106-180, 106th Congress, 17 March 2000.

Federal Communications Commission. Intelsat, Ltd. Files Petition for Declaratory Ruling and Certification Pursuant to Section 621(5)(F) of the Open-Market Reorganisation for the Betterment of International Telecommunications Act, as amended (the "ORBIT Act"). Pleading Cycle Established IB Docket No. 05-18. DA 05-88 (2005).

Constellation, LLC, et al., Consolidated Application for Authority to Transfer Control of PanAmSat Licensee Corp. and PanAmSat H-2 Licensee Corp., IB Docket No. 05-290, Memorandum Opinion and Order, FCC 06-85, 21 FCC Rcd 7368 (2006).

FCC Report to Congress As Required by the ORBIT Act. Eighth Report, of 12 June 2007. FCC 07-113.

# Chapter One

## Neo-Liberalism: A Contextual Framework for Assessing the Privatisation of Intergovernmental Satellite Organisations

*Patricia K. McCormick*

### 1. *Introduction*

The policy of reforming and privatising the telecommunications sector rapidly diffused on a global basis starting in the 1990s. While this development can be viewed from various contextual frameworks, perhaps the most informative is provided from the perspective of neo-liberalism. In this regard, the developments at the end of that decade relating to the privatisation and restructuring of the world's most prominent intergovernmental satellite organisations, namely, the International Telecommunications Satellite Organisation (INTELSAT), the International Mobile Satellite Organisation (INMARSAT) and, at a regional level, the European Telecommunication Satellite Organisation (EUTELSAT), would appear to be the natural extension of the tenets of neo-liberalism.[1] In the span of a few short years, these intergovernmental satellite organisations were essentially dismantled, as they were simultaneously transformed into private corporate enterprises, owned, no longer by governments and by extension, citizens, but now, in some cases, by private equity firms. This chapter seeks to assess these developments in the broader context of evolving tenets of neo-liberalism, including its affects on the global economy, developing states, and international institutions.

[1] Please note that henceforth in this document, INTELSAT, INMARSAT, and EUTELSAT shall refer to the intergovernmental satellite organisations prior to privatisation and Intelsat, Inmarsat, and Eutelsat will refer to the respective organisations and their subsidiaries after their privatisation.

## 2. *The Neo-Liberal Context*

The central premise of neo-liberalism, which has also been referred to as economic globalisation or corporate globalisation, is based on unfettered free trade and free markets. In essence neo-liberalism favours market-based mechanisms as preferred solutions to socio-economic problems, as it is "in the first instance a theory of political economic practices that proposes that human well-being can best be advanced by liberating individual entrepreneurial freedoms and skills within an institutional framework characterised by strong private property rights, free markets, and free trade."[2] Indeed, neo-liberalism values market exchange as an ethic in itself as it deems the political ideals of human dignity and individual freedom to be threatened, not only by fascism and dictatorships, but also by state interventions that substitute collective judgments for those of individuals free to choose.[3] Although individuals are supposedly free to choose, they are not supposed to choose to construct strong collective institutions, such as trade unions, or create political parties or instruments with the goal of state intervention in the market.[4]

A primary tenet of neo-liberalism is the assumption that individual freedoms are best assured by freedom of the market, derogating the concept of freedom into a mere advocacy of free enterprise.[5] This emphasis on market freedom serves as the principal feature distinguishing neo-liberalism from embedded liberalism, wherein market processes are entrenched in a larger context of social and political constraints as well as an accompanying regulatory environment that assists in meeting social needs. Amin argues that 'market' and 'capitalism' are two distinct concepts in that the 'market,' which, by its nature, refers to competition, is not the same as 'capitalism,' whose content is defined precisely by the limits to competition that monopolistic (or oligopolistic) private property entails, and that the existing confusion between the terms 'market economy' and 'capitalist economy' is at the root of a dangerous weakening of criticism of present neo-liberal policies.[6] Amin contends that capitalism does not function via

2 David Harvey, *A Brief History of Neoliberalism* (Oxford: Oxford University Press, 2005): 2.

3 Ibid., 5.

4 Ibid., 69.

5 Ibid., 7 and 183.

6 Samir Amin, *Obsolescent Capitalism: Contemporary Politics and Global Disorder* (London: Zed Books, 2003): 153.

competition among or between those who essentially hold monopoly power, but rather requires the intervention of a collective authority representing capital as a whole;[7] hence, the creation, through government-approved consolidation, of industries dominated by a small number of large, vertically integrated corporations. The state itself, thus, cannot be separated from capitalism or, for that matter, neo-liberalism.

### 2.1. *The 1980s – Global Economic Decline and Creditor Response*

The era of neo-liberalism was launched in the late 1970s and early 1980s with the election of Margaret Thatcher in Britain and Ronald Reagan in the United States. The Thatcher government, disregarding tremendous public resistance and rancour, pursued a vociferous and comprehensive program of denationalization and privatised many state industries.[8] The mainstream media presented the trade unions as greedy and their strikes disruptive, gradually weakening public sympathy and support for the unions.[9] The new doctrine of economic orthodoxy regulating public policy at the state level was concurrently embraced by the Reagan administration which too sought to curtail public sector trade union power as it promoted deregulation, which is essentially regulation in the corporate interest as it is regulation of markets by capital, increased market competition, and decreased government involvement and oversight of commercial enterprises. It also embraced the view that democracy and the market beget each other, in contrast to the view that holds that principles of capitalism, private property, market competition, and freedom of enterprise, have nothing to with democracy.[10] Further, institutions of global governance, deeply induced by the triad states of the United States, United Kingdom and Japan, and the complicated voting arrangements based largely on economic power, have also been complicit with the dominance of neo-liberalism.[11]

As neo-liberal policies found favour with the Washington-based Bretton Woods institutions, creating what has been termed the 'Washington

7 Ibid.

8 P.W. Singer, *Corporate Warriors* (Ithaca and London: Cornell University Press, 2003): 66–67.

9 Harvey, *Brief History of Neoliberalism*, 58.

10 Amin, *Obsolescent Capitalism*, 33.

11 Susanne Soederberg, *Global Governance in Question: Empire, Class and the New Common Sense in Managing North-South Relations* (London and Ann Arbor: Pluto Press, 2006): 29.

Consensus,' the world witnessed the momentous shift toward greater social inequality and the restoration of economic power to the upper class.[12] The lending practices of the World Bank and the International Monetary Fund (IMF), which are heavily influenced by the US Treasury, effectually disseminated neo-liberal policies, such as fiscal restraint, tax reform, trade liberalisation, deregulation, and privatisation of state enterprises, to developing countries as lender-imposed conditions. This conditionality essentially refers to the economic policy changes that external agencies imposed on national governments in the name of economic rationalization. In essence, conditionality transforms lending practices into policy articulation.[13] As indebted states were required to implement institutional reforms in exchange for debt rescheduling, the era of structural adjustment was ushered in.

When global economic growth slowed markedly in the early 1980s, many governments found it increasingly difficult to meet their external debt servicing requirements. For many developing countries, debt became the predominant policy issue of the decade, with attempts to meet repayment conditions effectively constraining policy options. The postulate that developing countries would come to finance an increasing proportion of investment from domestic savings had generally not been realised. Poor export performance coupled with the failure of economic diversification and the development of new foreign exchange earners resulted in a continued dependence on weak traditional sectors, and, thus, foreign savings. In the absence of commodity agreements that would assist in stabilizing prices, the significant drop in commodity prices in the 1980s further served to augment the deterioration of the terms of trade and thus the balance of payments position of many developing countries dependent on a few primary exports.

This fiscal crisis provoked a re-evaluation of governments' economic policies and also altered patterns of investment and multilateral lending, as, ironically, public institutions, such as the IMF, which was created to remedy market failures, came to be governed by market fundamentalists with little confidence in governments and associated public institutions.[14] The coalition of major governmental-based creditor entities, notably the US Treasury,

---

[12] Ibid., 26.

[13] Joseph E. Stiglitz, *Globalization and Its Discontents* (New York and London: W.W. Norton & Company, 2002): 44.

[14] Ibid., 196.

the World Bank, and the IMF, accompanied by the major global commercial banks, assumed a pivotal role in the design of economic reform programs, focusing on the internal or domestic elements affecting an economy.

In response to these perceived problems, the programs of the external agencies, in accord with neo-liberalism, adopted the dominant orthodox view of development, stressing export-led growth with internal price liberalisation and reduced direct government intervention in the economy, a minimalist state, with the major instrument of reform being the market. To stimulate growth on a sustainable basis, developing countries, in the view of both multilateral and bilateral agencies, needed to employ not only improved macroeconomic management and price reforms, but to restructure public investment and state owned enterprises.

The World Bank stressed the virtues of market-led reform when it introduced Structural Adjustment Loans (SALs) in 1980, later supplemented by Sector Adjustment Loans (SECALs), thus, shifting a portion of its traditional project lending into loans conditioned on policy and institutional reform. The conditionalities of IMF agreements were also further broadened. Indeed, the 1980s witnessed unprecedented external intervention in the internal economic policies of an unparalleled number of countries in terms of the measures addressed and conditionalities and cross-conditionalities ascribed. Cross-conditionality describes a situation in which a country is compelled to accept the conditions of one Bretton Woods institution in order to qualify for assistance from the other. Cross-conditionality, thereby, enables the multilateral lending agencies to exercise greater control over a country's domestic economic management. Furthermore, the IMF imprimatur or seal of approval is required by commercial banks and bilateral agencies engaged with debtor nations, which are often compelled to borrow more in order to maintain positive net inflows and are therefore exposed to extended and enlarged conditionality.

Neo-liberalism arguably thus runs counter to certain national democratic processes. Regardless of elections and changes in heads of state, governments are locked into a series of reform policies as power shifts from centralized control of nation-states to centralized control of global corporations and global institutions.[15] Due to debt and debt servicing requirements, developing countries are essentially forced into these

[15] Vandana Shiva, *Earth Democracy:Justice, Sustainability, and Peace* (Cambridge, MA: South End Press, 2005): 6 and 88.

externally designed reform programs, though not because of any conviction of their inherent worth. It is not simply the policies themselves that developing states oppose, but the assumption that there is but a single set of policies that is right for each and every state.[16] Further, when countries view reforms as being imposed on them, they fail to invest and commit to the reforms, though full participation is essential if societal transformation is to occur.

Indeed, the resumption of growth with economic stability and a flexible and adequate response to external shocks, including a decreased vulnerability to changes in the international market, have proven formidable goals to achieve. Aims of the restructuring programs, including revision of the banking sector; reductions in the role of the state; institutional and public sector reforms, which may entail divestment; and price reforms, such as reducing or eliminating subsidies for state owned enterprises, essentially present an orthodox paradox, since governments are now to adopt policies that may be economically rational, but are politically irrational as the state is to dismantle its own power. States are forced to sacrifice their sovereignty to capricious capital markets, including spectators concerned solely with short-term profit rather than the long-term economic growth of the country and the improvement of society.[17]

Development emphasises the orientation of production towards basic needs, the orientation of distribution and services towards the poor majority, the incorporation of a social dimension into technological research and innovation, and popular participation in a democratic system. The goal of development is the improvement of society through strong democratic practices and institutions and economic growth, accompanied by full employment and income distribution favouring the popular classes, objectives not clearly identified by the Bretton Woods institutions as part of their agendas. The original mandate of the IMF was to engender global economic stability, not reduce poverty in developing countries.[18]

Development of the periphery is not a goal of capital's strategies, for not only does capitalism reduce citizens to consumers, but dominant capital also finds it profitable to manage the Third World debt.[19] In 1982, the IMF

---

16 Stiglitz, *Globalization and Its Discontents*, 221.

17 Stiglitz, *Globalisation and Its Discontents*, 247.

18 Ibid., 34.

19 Samir Amin, *Capitalism in the Age of Globalisation: The Management of Contemporary Society* (London: Zed Books, 1997): 34 and 40.

and the World Bank were given full authority to negotiate debt relief, in effect, protecting the world's main financial institutions from the threat of default, though in accord with neo-liberal theory, investors should be responsible for their own mistakes, thus making any protective measures undertaken by the Bretton Woods institutions unnecessary.[20] Debt crises have grown in frequency since the 1980s, and their orchestration has proven a means to both rationalise the system and redistribute assets.

Neo-liberalism and its accompanying views toward increased international trade and investment as the route by which all countries can most effectively improve standards of living have been exported across the world. This ideology of market fundamentalism requires little, if any, consideration of a given country's specific historical circumstances and current quandaries.[21] Although neo-liberalism, as noted, constitutes several fundamental components or policy proposals, notable among them for the purposes of this book, is the privatisation and commodification of public services and aspects of the global commons, including space and space assets. Neo-liberalism is particularly assiduous in seeking the privatisation of assets, for the absence of clear private property rights is seen as an institutional barrier to economic development.[22] The corporatization, commodification, and privatisation of hitherto public assets is a key feature demarking neo-liberalism, which it justifies in its contention that enclosure and the assignment of private property rights is the best means to protect against the so-called 'tragedy of the commons,' that is, the tendency for individuals to irresponsibly super-exploit common property resources such as land and water.[23]

Lost in the tenets of neo-liberalism is the view of the commons as common heritage resources, a collective birthright or inheritance, if you will, that should be equally shared since they support diverse life forms on the planet, as exemplified by water, land, air, forests, and fisheries.[24] Some commons may be thought of as global, such as the atmosphere, the oceans, and space because they are beyond sovereign state jurisdiction and have no territorial claimants, while others may be considered as community

20 Harvey, *Brief History of Neoliberalism*, 73.

21 Stiglitz, *Globalization and Its Discontents*, 36.

22 Harvey, *Brief History of Neoliberalism*, 65.

23 Ibid., 160 and 65.

24 John Cavanagh et al., *Alternatives to Economic Globalization: A Report of The International Forum on Globalization* (San Francisco: Berrett-Koehler Publishers, Inc., 2002): 19.

commons, such as public spaces, common lands, the gene pool, and the culture and knowledge that are collective creations, such as local knowledge with respect to medicinal plants.[25] More contemporary use of the term commons includes public services that governments perform to address basic needs such as water systems, public health care, education, and public safety, among others.[26] The broadcast spectrum and the Internet also arguably constitute the commons. Although the use of a common heritage resource carries a corresponding obligation to act as its steward, the commons are under tremendous strain as the thrust of neo-liberalism has been driven by the efforts of corporations to commodify and market these resources. Water, for example, a common heritage resource essential to the survival of all species, has come to be a tradable commodity as states and municipalities privatise water and water services. The World Trade Organisation (WTO) further specifies provisions prohibiting the use of export controls to prevent the exportation of water.[27]

Governments, from a moral perspective, should arguably serve as trustees of the commons and not authorise their enclosure or privatisation, but this path has been embraced as part of the ideology of corporate globalisation.[28] While Shiva notes five processes as constituting the enclosure of the commons, notable among them is the creation of private property by the enclosure of common property.[29] A privatised commons is in fact no longer a commons, it is private property, either *de facto* or *de jure*.[30] Indeed, it may be argued that public-private partnerships are essentially private appropriations of public investment, since public taxes initially funded many industries that were later privatised, including, for example, the development of telecommunications systems in many states and the Internet, which was initially funded through the US Department of Defense as the Advanced Research Projects Agency Network (ARPANET).[31] Similarly, INTELSAT, as a monopoly service provider, not only benefited from US taxpayer-funded research and development conducted in the pioneering days of space communications, but also US government policies designed

---

[25] Ibid., 63 and 81.

[26] Ibid., 63.

[27] Ibid., 84.

[28] Shiva, *Earth Democracy*, 2.

[29] Ibid., 20.

[30] Ibid., 55.

[31] Ibid., 173.

to assure its commercial success so as to ensure the attainment of broader public policy goals, goals generally lost in the process of privatisation.[32]

2.2. *Privatisation*

The proffered rationale favouring the privatisation and institutional reform of state owned enterprises may be seen as equally applicable to intergovernmental enterprises, such as the intergovernmental satellite organisations herein under examination, regardless of whether private companies or governmental agencies of the countries party to the intergovernmental agreement or a combination thereof comprise the entity in question. Institutional reform of state owned enterprises due to perceived weak economic performance has been an integral component of structural adjustment programs. The World Bank has suggested that when state owned enterprises lose sight of profitability, both economic and social objectives may be sacrificed.[33] Profitability, however, may not be the best indicator or measurement of performance as it may well be a peripheral concern since the public enterprise is often an instrument for political patronage.

The principal economic issue in the privatisation debate concerns efficiency, one aspect of which concerns pricing, which is to be brought in line with costs, thereby ending cross-subsidization.[34] However, for reasons of more equitable distribution, public enterprises might assign low prices, which would thus result in lower profitability. This is arguably somewhat analogous to INTELSAT's mandate to charge uniform prices for similar services, which was considered a subsidy to the thin-route users in developing countries because prices were not related to the actual costs of service provision. The overall deficits of state owned enterprises, though admittedly an inadequate and perhaps even a misleading indicator of performance have, however, prompted varying measures of institutional reform. The inefficiency of the public sector, of which intergovernmental institutions

32 Linda L. Haller and Melvin S. Sakazaki, "Commercial Space and United States National Security" (paper prepared for the Commission to Assess United States National Security Space Management and Organisation Washington, D.C., 2000): 9.

33 Don Babai, "The World Bank and the IMF: Rolling Back the State or Backing its Role," in *The Promise of Privatisation: A Challenge for US Policy*, ed. Raymond Vernon (New York: Council on Foreign Relations, Inc., 1988): 264.

34 Francesco Stolfi and Gerald Sussman, "Telecommunications and Transnationalism: The Polarization of Social Space," *The Information Society* 17 (2001): 51.

are part, is generally associated with bureaucratic failure and political interference, including political patronage. It is thus argued that privatisation, by removing these obstructions, will improve a firm's efficiency.

Though state enterprises, properly managed, can play a vital role in economic transformation, privatisation is perhaps an easier option to implement than other forms of institutional restructuring. Reallocating and retraining personnel and altering incentives for managers of public enterprises so as to encourage new forms of behaviour, including more autonomous decision-making and increased accountability, appear to be beyond the capabilities of many institutions. Part of the problem is that such institutional reforms require time to work effectively, whereas a divestment can often be concluded fairly quickly.

Indeed, the activity with which privatisation has become most closely associated is divestiture or the sale of public sector assets, though privatisation can take various forms, including management contracting and leasing or franchising, which may effectively remove the government from direct control of the assets through actual ownership remains unchanged. Competition can enter at the stage of bidding for a franchise. Liberalisation or deregulation of entry into activities previously restricted to the public sector, too, does not entail a transfer of ownership assets, though it does present opportunities for private, and often foreign, investors.

While some contend that management, not ownership, is the key to efficiency of an enterprise, the property rights school suggests that privatisation in the form of a change in ownership, an alteration of the structures of property rights, will improve the incentives for productive efficiency performance.[35] It recommends the privatisation of public enterprises operating in competitive markets for competition compels companies to improve their performance, that is, increase allocative and productive efficiency in order to profit. It would thus seem that an improvement in the economic performance of the public enterprise sector is more likely to result from an increase in market competition than from a change in ownership.[36]

In relation to satellite operations, President Ronald Reagan, in accord with his administration's neo-liberal policies promoting deregulation,

---

35 Colin Kirkpatrick, "Privatisation in Less Developed Countries: An Overview," in *Privatisation in Less Developed Countries*, eds. P. Cook and C. Kirkpatrick (Brighton: Wheatsheaf Books, Ltd., 1988): 19.

36 Ibid., 31.

increased market competition, and less government involvement in commercial enterprises, determined on 28 November 1984 that separate international satellite systems were required in the country's national interest. Orion and the Pan-American Satellite Corporation (PanAmSat), in addition to four other companies, made their initial applications to the Federal Communications Commission (FCC), and, despite opposition from INTELSAT, the FCC proceeded to license these systems.[37] Indeed, with the launch of PAS-1 in 1988, PanAmSat (PAS) became the first private satellite operator to directly compete with INTELSAT and attempt to break its monopoly.[38]

The natural monopoly argument, which contended it was more economically efficient to construct, in this case, a single satellite network connecting different countries as opposed to building several competing satellite systems, gave rise to the formation of INTELSAT as a single monopoly provider of satellite services. This prevailing economic rationale, which held that natural monopolies in the provision of public utilities were the most efficient, supported the telecommunications structure historically adopted in virtually every country in the world, in that providers of telephony and telegraph services were either private monopolies, crown corporations, or, most commonly, government Ministries of Post, Telegraph and Telephone (PTTs). The large initial costs or high sunk costs in building and operating networks, and similarly, satellites, coupled with the relatively low cost of adding additional users in conjunction with the arguments about economies of scale and unnecessary duplication of costly facilities, contributed to the general assumption that competition in telecommunications and satellite communications was neither feasible nor desirable.[39]

Although privatisation in the form of divestiture in itself does not lead to the creation of a competitive environment or to increased entrepreneurship, it is argued that the replacement of a public monopoly by a private monopoly will increase productive efficiency due to the impact of reduced political interference and more effective financial constraints. Forcing

---

37 Robert W. Hahn and Randall S. Kroszner, "Lost in Space US International Satellite Communications Policy," *Regulation The CATO Review of Business and Government* 13, no. 2 (Summer 1990): 62.

38 Johannes P. Pfeufenberger and Hendrick S. Houthakker,"Competition to International Satellite Communications Services," *Information Economics and Policy* 10 (1998): 408.

39 Hahn and Kroszner, "Lost in Space US International Satellite Communications Policy," 58.

privatisation before adequate competition and regulatory frameworks are in place can, however, incur real costs, especially in developing countries which are ill-equipped to handle these institutional transformations.[40] This mistake in sequencing and pacing of the privatisation process arguably only benefits the foreign entrepreneurs who purchased the enterprise and upon whom the developing state is now dependent.[41] To entice foreign investors, however, some developing countries, notably those who entered the privatisation process late, permitted their newly privatised telephone companies to retain a service monopoly for a period of several years.[42] In some instances, privatised monopolies, without regulatory mechanisms in place, proved to be more capable of exploiting consumers than the state monopolies.[43] Without the essential institutional regulatory infrastructure, privatisation can lead to asset stripping rather than wealth creation.[44]

Analysis of privatisation cannot be confined to debates concerning economic efficiency, but must address the larger political context for such decisions. In a national context, many state owned enterprises constitute an agency of employment. Fears that privatisation will increase unemployment may inhibit the political regime from engaging in divestment for such a decision could entail the loss of support from both public sector employees and their representative trade unions. Privatisation thus should be part of a comprehensive program, which focuses on job creation in conjunction with the inevitable job destruction that privatisation often entails.[45] Although employee share schemes aim at diffusing opposition to privatisation, and coupled with public share offerings, assist in broadening the structure of private ownership, such methods of divestment are generally constrained in developing countries by the thinness of capital markets and the embryonic stature of the stock exchanges.

In such instances, unless efforts are made to increase market capacity to handle the float, denationalization is more likely to involve the sale of the enterprises as a complete entity, or, at least, controlling interest is sold to a single, often foreign, buyer. Inadequate levels of domestic savings coupled

40 Stiglitz, *Globalization and Its Discontents*, 54 and 73.

41 Ibid., 72.

42 Dan Schiller, *Digital Capitalism: Networking the Global Market System* (Cambridge, MA: The Massachusetts Institute of Technology Press, 2000): 65.

43 Stiglitz, *Globalization and Its Discontents*, 220.

44 Ibid.

45 Ibid., 57.

with capital flight in the context of stagnant or negative growth rates has forced many governments to adopt a more favourable disposition to private foreign investment. Even if foreign ownership is deemed politically unacceptable and widening international inequalities in the distribution of wealth, a government may still find it necessary or desirable to sell to foreign interests if it is sold to an entity engaged in a similar activity. In the telecommunications sector, for example, governments of developing countries typically seek partners with considerable experience in building and operating networks.

Although partners are coveted for their technological and management capabilities, financial commitments are equally important, for many governments find themselves faced with the dilemma of replacing antiquated systems with modern equipment at a capital cost beyond their reach. It is thought that private investors with access to commercial lending sources and expanded credit possibilities will be better able to provide the capital necessary for the development of the telecommunications infrastructure and diverse services than a government which faces competing claims on funds raised by taxation. Privatisation is also a means to acquire revenue for other programs and debt reduction.

The shortage of foreign exchange has been a compelling motive for many privatisation programs. To scale back the burgeoning state sector by selling salable public assets to the private sector has been considered by some governments as a means of handling an acute capital shortage and acquiring funds for debt servicing or capital investments. For countries whose deficits and debts have grown beyond control and cannot be reversed by a continuation of the policy of state ownership, divestiture has become viewed as a means to raise revenue and reduce fiscal and credit pressures, that is, reduce the high levels of borrowing, in part, by ending subsidies.

If the returns from a sale are applied strictly towards external or domestic debt, the effect will be but a one-time reduction in the government deficit, equal to the amount of the sales revenue. If the public enterprise is profitable, hence a more attractive candidate for sale, its privatisation means that the government forfeits the future stream of income, unless it retains a large share, a partial divestiture. The sale of profitable firms could, however, be made on the basis that proceeds from the sale would maintain social services and finance faster growth. There are several options open to governments in the use of privatisation proceeds. These options include using the revenue for current expenditures; tax reductions; 'social' capital expenditures; 'commercial' capital expenditures, including the restructuring of selected public enterprises; financial private investments; and, as noted, public debt reduction or non-increase in public borrowing. Regardless of the revenue application, it is imperative that a regulatory structure be established prior to the divestiture, though in many instances, this does not occur.

### 2.3. *The Role of Regulation*

Regulation may be defined as the substitution of rules made by government for the competition of the market. Administrative law, policy, and practice apply regulation and similar terms somewhat differently in every country. Typically, regulation means control of some social authority by a duly authorised administrative agency of a government, though courts too possess powers of administration.

When the state or an intergovernmental body sells its assets, it does not necessarily abandon its regulatory or arbitrator function. In fact, it can come to have an even greater responsibility in establishing the rules of operation, for ensuring that the rules are adhered to, and for sanctioning transgressions, as private operators adhere to market logic in contrast to the previous public service ethic. Thus, as states reduce their operational role and ownership stake in a given sector, such as telecommunications and/or satellite services, the biggest challenge is to define or in some instances re-define the proper regulatory role that the state is to play in order to effectively accomplish certain policy objectives in the absence of public ownership.

When the private sector plays a significant part in the provision of telecommunications and/or satellite services, there are three essential roles that the government(s) may need to perform: (1) award and regulate franchises; (2) specify appropriate technical standards; and (3) ensure access to all systems.[46] The International Telecommunication Union (ITU) has identified several additional regulatory policy issues. These include the need to: (1) define the distinction between public and private services, (2) interpret the law and reconcile policy objectives, (3) ensure fair competition for new entrants, (4) ensure efficient procedures for interconnection between new and existing service providers, (5) verify reasonable pricing to cost rations

[46] Gabriel Roth, *The Private Provision of Public Services in Developing Countries* (Oxford: Oxford University Press, 1987): 188.

and relate this to quality of service, (6) authorise and assure transparency of schemes for subsidies where required, and (7) establish clear-cut dispute resolution procedures.[47] Regulation can thus set the conditions for the sector's operations, the rules for entry, the extent of competition, exit guarantees (in the case of foreign investments in the sector), type approval, general surveillance, pricing of monopoly services and service quality, and legal procedures for conflict resolution.[48]

Regulation is arguably intended to effect a transfer of wealth from producers to consumers and may be regarded as 'social' when it safeguards those interests which the market cannot be expected to meet through the operation of the profit motive.[49] Profit-based telecommunication or satellite entities cannot be expected to voluntarily finance uneconomic activities, such as the provision of service in rural or remote areas. If the goal is universal service, committed government policy is essential. A purely market-driven system of allocation will tend to produce a telecommunications network that concentrates disproportionately in the main cities and on the largest and wealthiest customers. Price and profit controls are required to prevent a monopoly supplier of services from extracting excessive profits from the business.[50] Profit ceilings and ways to make companies invest in non-business areas as well as cross-subsidies to provide universal service can be required of private monopolies through regulation.

It must be recognised that the transfer of ownership of a public utilities monopoly to the private sector should require some degree of regulation to ensure social efficiency through fair pricing and other arrangements, since privately owned telecommunication or satellite entities may be tempted to use their relative or absolute monopoly position to maximize profits by increasing charges to unreasonable levels or by reducing expenditures for service quality and expansion. These profit-maximizing tendencies require

---

47 International Telecommunication Union Centre for Telecommunications Development, "Restructuring of Telecommunications in Developing Countries" (Geneva: ITU, April 1991), Presented at the Seminar on Implementing Reforms in the Telecommunications Sector: Lessons from Recent Experience, Organised by the World Bank, the International Telecommunication Union, the Commonwealth Telecommunications Organisation, and the Centre for Telecommunications Development, Washington, D.C. (23–26 April 1991): 25.

48 Ibid., 18.

49 Jill Hills, *Deregulating Telecoms: Competition and Control in the United States, Japan and Britain* (Connecticut: Quorum Books, 1986): 29.

50 Bjorn Wellenius et al., eds., *Restructuring and Managing the Telecommunications Sector* (Washington, D.C.: World Bank, 1989): 12.

direct government regulation of utility charges and service quality, modified where possible by the introduction of some form of competitive pressure.[51] Some government or intergovernmental agency must be prepared to effectively regulate, if not remove or reduce, the private monopoly if service is unacceptable or rates unreasonable. Indeed, despite competition in satellite services, such regulatory arrangements were created for Intelsat, Inmarsat, and Eutelsat.

The privatisation of the three intergovernmental satellite organisations was structurally similar in that intergovernmental agencies were retained to ensure that the private companies, to which all assets and activities were respectively transferred, adhere to public service standards. In regards to Intelsat, a continuing intergovernmental organisation, retaining the name the International Telecommunications Satellite Organisation with the acronym 'ITSO,'[52] was designated to serve in a supervisory and monitoring capacity, ensuring that the private company meet its public service and Lifetime Connectivity Obligations (LCO) to those developing countries largely dependent on the services and pricing mechanisms of Intelsat, Ltd. LCO protection will phase out beginning in July 2013. In the case of INMARSAT, the intergovernmental body, the International Mobile Satellite Organisation (IMSO),[53] was established to ensure that INMARSAT continues to meet its public service obligations, including those relating to the Global Maritime Distress and Safety System (GMDSS). Similarly, EUTELSAT IGO[54] was created to ensure that Eutelsat adheres to the principles of pan-European coverage, fair competition, and universal service. The roles of each of these organisations, however, were narrowly proscribed and they were not granted more broad-based regulatory authority.

While regulation of prices may be needed if there is only one supplier in the market, when there are multiple suppliers, there is arguably no need for price controls, assuming that the market structure is not dominated by a single firm or oligopolistic cartel and that competition is workable.

---

[51] Ibid., 34.

[52] Agreement Relating to the International Telecommunications Organisation, done 20 August 1971, entered into force 12 Feb. 1973, as amended 17 Nov. 2000, applied provisionally 18 July 2001, entered into force 30 Nov. 2004, 23 U.S.T. 3813 and 4091, T.I.A.S. 7532.

[53] Convention on the International Mobile Satellite Organisation, London, entered into force 16 July 1979, as amended 1998, amended version applied provisionally 15 April 1999, entered into force 31 July 2001, 31 U.S.T. 1, T.I.A.S. 9605.

[54] Convention Establishing the European Telecommunications Satellite Organisation, done 15 July 1982, entered into force 1 Sept. 1985, as amended 20 May 1999, applied provisionally 2 July 2001, entered into force 28 November 2002, UK Misc. No. 4, Cmnd. 9154.

Pro-competitive regulation can use the attempts by market entrants to compete or to find profitable niches in the market, and serve as a check on the pricing, quality, and efficiency of the incumbent entity, thus obviating to some degree the need for price and profit controls.[55] Some intergovernmental agency or government body with specific legal authority, however, must oversee, coordinate, and enforce uniform standards of performance for transmission and terminal equipment when the system includes more than one operating entity.[56]

In sum, as the telecommunications and satellite industries yield to more complex structures and growing number of participants, the government's role in encouraging and regulating sector activity becomes distinct and increasingly important.[57] Governments need to establish policy-making and regulatory capabilities separate from the operating entities and not subject to undue political influence. An effective regulatory agency should have a considerable degree of financial and administrative autonomy. It should not only be essentially independent in its regulatory decisions, but invested with sufficient autonomy to limit the possibility of it being captured by particular interest groups. Although the utilization of regulatory power can allow governments to divest state owned enterprises while retaining some control over service provision, pricing, and the direction of the industry, these efforts may be restrained by the larger structures of global governance.

In a neo-liberal environment, regulation tends to be dismissed as protectionism or an infringement of market freedom,[58] since neo-liberalism embraces deregulation or a shift to market-driven regulation. This tendency has been manifested in the field of telecommunications in various arenas, including the WTO and the accompanying General Agreement on Trade in Services (GATS), as well as within the ITU itself.

## 2.4. *Pressures for Deregulation: WTO and GATS*

The ascendance of market logic or market fundamentalism has created a transnational managerial class of dominant states and their corporations

55 International Telecommunication Union Centre for Telecommunications Development, "Restructuring of Telecommunications in Developing Countries," 23.

56 Wellenius et. al., *Restructuring and Managing the Telecommunications Sector*, 31.

57 Ibid., 94.

58 Cavanagh and Mander, *Alternatives to Economic Globalization*, 212.

who have created a globally hegemonic presence promoting neo-liberal reform policies in such multilateral institutions as the World Economic Forum at Davos and the WTO.[59] While there is indisputably a need for a multilateral system of rules regarding international trade and finance so as to ensure stability, predictability, and fairness for all participants, the current system of global governance fails to provide such assurances as it represents only the narrow financial self-interests of the global elite.[60] Indeed, as noted, analysis of privatisation cannot be confined to debates concerning economic efficiency, but must address the larger political context for such decisions and the ascendance of what Cox terms the transnational managerial class of dominant states and their corporations who have created a globally hegemonic presence promoting neoliberal reform policies in such multilateral institutions as the WTO.[61]

Shiva contends that 1995 demarcates the year when corporate globalisation became the legal constitution of the world for it was the year in which the WTO was established, essentially as the third pillar of the Bretton Woods institutions, imposing a supranational global trade architecture whose primary beneficiary was transnational capital.[62] The opening of capital markets is now a condition of membership of the IMF and the WTO, drawing developing states into the neo-liberal fold.[63] It is further arguable, from a political, if not inherently legal, perspective that these intergovernmental institutions, as well as national governmental bodies, function as if sovereignty resided in global corporations.[64] The absence of transparency in negotiations that result in agreements makes it difficult, until it is too late, to recognise the influence of corporate and other special interests.[65] Moreover, the connectivity between corporations and financial markets or stock exchanges increased during the 1990s as financial markets experienced a powerful wave of innovation and deregulation internationally.[66] Financial markets became both critical instruments of co-ordination and

59 Robert W. Cox, *Production, Power and World Order: Social Forces in the Making of History* (New York: Columbia University Press, 1987): 359–360.

60 Cavanagh and Mander, *Alternatives to Economic Globalization*, 216 and 224.

61 Cox, *Production, Power and World Order*, 359–360.

62 Shiva, *Earth Democracy*, 31.

63 Harvey, *A Brief History of Neoliberalism*, 72.

64 Cavanagh and Mander, *Alternatives to Economic Globalization*, 140.

65 Stiglitz, *Globalization and Its Discontent*, 227.

66 Harvey, *A Brief History of Neoliberalism*, 90.

provided the means to procure and concentrate wealth, in essence, the privileged means for the restoration of class power, a key goal, albeit unstated, of neo-liberalism.[67] The WTO, coupled with the structural adjustment policies of the World Bank and the IMF, effectively opened up domestic markets in developing countries for transnational corporations to enter as it opposed protectionism in the areas of trade and finance and encouraged privatisation of sectors previously considered "off limits," such as water, electricity, transportation, banking, and telecommunications.[68]

The principle historic step toward creating a more competitive commercial global telecommunications and satellite market was the 1997 World Trade Organisation Protocol to the General Agreement on Trade in Services.[69] The sixty-nine countries that were initial signatories to the WTO Agreement, accounting for more than ninety percent of the world's basic telecommunications revenue, agreed, among other commitments, to provide unconditional market access to all international telecommunications services and facilities.[70] This binding agreement also empowered the WTO to authorise penalties for those signatory countries that were not in compliance. In limiting the ability of national governments to regulate their own activities, a core principal of democracy was effectively undermined – the right of a society to legislate for itself and thereby construct its own future. Thus, as characterised by Amin, "the WTO deserves to be called the 'colonial ministry' of the collective imperialist."[71]

Certainly, this extraterritorial corporate charter bore far-reaching consequences for national telecommunications provision, particularly in developing states, as this marked shift from a state-centric to a market-oriented view of communications among major global powers eroded the political and market authority of PTT entities as countries agreed to permit competition in telecommunications services, including non-discriminatory market access to satellite markets. Indeed, as part of this shift, effective on 1 January 1998, by WTO decree, the world's skies opened to satellite competition, but without adequately addressing equitable management and

---

67 Ibid.

68 Shiva, *Earth Democracy*, 69 and 70.

69 General Agreement on Tariffs and Trade ATS 1995 No. 8, Apr. 15, 1994, 55 U.N.T.S. 194, entered into force 1 Jan. 1995.

70 Haller and Sakazaki, "Commercial Space and United States National Security," 50.

71 Amin, *Obsolescent Capitalism*, 116.

development of 'common heritage' principles as related to space.[72] Since space and the electromagnetic spectrum are part of our global commons, intergovernmental satellite organisations as well as their private successors can be defined as common property resource regimes; however, a founding legal principle of the capitalist system is the sanctity of property, hence, the emphasis on privatisation.[73] Markets, however, cannot produce equilibrium, thus capitalism is synonymous with permanent instability.[74] Regulation, while imperative, is often insufficient to safeguard public services, as evident in global telecommunications carriers obtaining commitments allowing foreign investment in existing national public service providers, enabling them to acquire, establish, or hold significant stakes in telecommunications networks worldwide, thereby facilitating transnational corporate expansion.[75]

Further incursions on national sovereignty were ensured by capital's guardianship of the Internet, as neo-liberalism required the development of supranational systems and applications of information technologies to meet the needs of corporations whose offices, factories, and outsourced facilities increasingly spanned the globe.[76] To best facilitate these corporate information requirements, telecommunications structures and policies of long standing were fundamentally revised so to "release the forces of the market" by eliminating existing state monopolies in telecommunications through their privatisation.[77] Corporate users were also freed to develop their own networks and services, independent from the public telecommunications system. As public service policies governing this critical infrastructure were replaced by market driven tenets, disparities in access widened as market mechanisms privileged those with the resources to afford high-tech Internet and mobile applications.[78] Similarly, in a democracy, public service is at least potentially open to public scrutiny whereas information pertaining to services provided by private firms is protected as proprietary.[79] Intellectual property rights are key legal instruments underpinning power and wealth, and since 1995 the Trade-Related Aspects of

72 Peter Speigel, "Lost in Space" *Forbes* 160, no. 6 (1997): 123.
73 Amin, *Obsolescent Capitalism*, 48.
74 Ibid., 23.
75 Schiller, *Digital Capitalism*, 47.
76 Ibid., 88 and 40.
77 Ibid., 44.
78 Ibid., 2 and 203.
79 Amin, *Obsolescent Capitalism*, 38.

Intellectual Property Rights (TRIPS) agreement has been overseen by the WTO.[80]

This is in accord with what Geri refers to as the New Public Management, in which the state, devoted to securing conditions that enable the domination of multinational firms, engages in administrative reforms which include divestment of state owned enterprises, deregulation, competition in services, and facilitation of the role of the private sector in international or intergovernmental organisations, if not, their actual privatisation.[81] To further a broad set of collective industry interests, as opposed to individual companies seeking preferential treatment, businesses have gradually refined their ability to act as a group, submerging competitive behaviour in favour of joint, cooperative action in the legislative arena.[82] In Washington, corporate lobbyists effectively dictate legislation in accord with their special interests, as they benefit from the revolving door between state employment and the immensely more lucrative corporate employment.[83] The revolving door between government and private sector employment also corresponds with capture theory, which contends that policy makers, such as FCC Commissioners, are effectively captured by special interest groups or industry, and because they identify with industry interests, they thus regulate to protect those interests rather than the "public interest." In sum, this political strategy of businesses or corporations is evident in the immeasurable influence they extend to mould and shape the policies of both national institutions, such as in the United States, the FCC and Congress, and international regimes, such as the WTO and the ITU.

### 2.5. *Pressures for Deregulation: ITU Restructuring*

Since the organisational structure of international telecommunications regulation, specifically that of the ITU, the United Nations agency which coordinates satellite spacing and oversees access to the electromagnetic

---

80 Christopher May, "Intellectual Property Rights, Capacity Building, and "Informational Development" in Developing Countries," in *Governing Global Electronic Networks*, eds. William J. Drake and Ernest J. Wilson III (Cambridge, MA: The Massachusetts Institute of Technology Press, 2008): 401.

81 Laurance R. Geri, "New Public Management and the Reform of International Organisations," *International Review of Administrative Sciences* 67, no. 3 (2001): 445.

82 Harvey, *A Brief History of Neoliberalism*, 48.

83 Ibid., 77.

spectrum on an international basis, reflected the preference for national sovereignty that typified a prior age, it was essential that it too be restructured to promote the neo-liberal agenda.[84]

Although US delegates to the ITU are charged with representing US policy, and not any particular company or private interest, this is to a large extent immaterial given the crucial role of the private sector in initially formulating the policy, not only in the United States but in other developed countries as well. Indeed, the vast majority of private sector delegates to the Plenipotentiary Conference, the supreme organ of the ITU, come from the core countries that dominate the global telecommunication and satellite markets, namely, the United States, the United Kingdom, Japan, Canada, France and Italy, and, given the decentralized structure of conference proceedings, the influence of these delegates far exceed their actual numbers.[85]

The private sector has extended its influence in the three Sectors of the ITU which correspond to its main areas of activity: Telecommunication Standardization (ITU-T), Radiocommunication (ITU-R), and Telecommunications Development (ITU-D). Although the ITU is still referred to as an inter-governmental organisation, since it comprises almost all of the world's countries which are designated as Members who alone retain full voting rights, it now encompasses more than 650 private companies from the telecommunication, broadcasting, and information technology sectors that are classified as Sector Members. The 1998 and 2002 Plenipotentiary Conferences focused on strengthening the participation of the private sector in the ITU, adopting several resolutions enhancing the rights of Sector Members, as well as measures to enable the ITU to match industry's time-frames and operational practices. Private Sector Members were also admitted on a provisional basis as observers at Council 2005 and 2006 sessions. Some contend that such progress in expanding the rights of the private sector has been meagre, given the amount of effort extended over many years to increase their influence, and many segments of the global telecommunications industry, who could provide valuable expertise, still consider participating in the ITU to be irrelevant and unnecessary for expanding their operations as this can be attained through bilateral negotiations. Others, however, aver that in light of the increased private sector

---

84 Schiller, *Digital Capitalism*, 40.

85 Kelley Lee, *Global Telecommunications Regulation: A Political Economy Perspective* (London: Pinter, 1995): 90.

participation in ITU activities it is essential that the needs of developing countries remain at the forefront of the ITU agenda and do not succumb to corporate directives.[86]

Like many nation states, though, the ITU has suffered notable financial duress in recent years, and funding issues are arguably part of the impetus for reform. In this regard, Sector Members have certainly assisted in alleviating the ITU's budgetary woes, which accounts in part for the organisation's campaign to recruit private companies to become active members and fund its various programs. It need be noted, though, that threatening to reduce one's financial contribution can also be employed as a means to change an organisation, or at least as a limited retaliation or demonstrated opposition to an organisational decision, as demonstrated by some European firms which reduced their ITU contributions when voting rights were denied to Sector Members.

According to the ITU, Sector Members contribute 11.6 percent and Associate Members, a more limited class of private company participants, contribute .6 percent of the overall budget. Notably, other forms of private sector financial contributions account for a further 15.2 percent of total funding, namely, cost recovery for services such as project execution, satellite notifications, and the sale of publications, thus raising total private sector contributions to more than 27 percent of the ITU budget. It should also be noted that as a group Sector Members and Associates pay about as much as that paid in total by approximately 88 percent of the Member States. This is due to the large number of developing Member states. Furthermore, if a Sector Member joins all three Sectors, that Sector Member pays nearly the amount that 60 percent of the Member States pay, again a situation which is attributed to the lower amounts paid by developing countries.[87] The most noteworthy change within the ITU regime, though, concerns the Telecommunication Standardization Sector (ITU-T), formerly the International Telephone and Telegraph Consultative Committee (CCITT), where Sector Members have come to play an equal role with governments in determining the agenda for each standardization cycle.

Study groups of experts from the private sector, that is, the leading international telecommunication organisations, conduct the technical work of

86 Patricia McCormick, "Private sector influence in the International Telecommunication Union," *info* 9, no. 4 (2007): 70.

87 "ITU Budget 2006–2007," International Telecommunication Union, accessed February 23, 2013. Available: http://www.itu.int/aboutitu/budget/2006-2007/index.html.

developing standards and specifications for telecommunications systems, networks and services. In few other international industries are technical interconnection standards of more importance than in telecommunications. Until 1989, telecommunications equipment standards were approved only every four years as recommendations by the plenary assemblies of the former CCITT and the International Radio Consultative Committee (CCIR), which was an extremely long time frame given the rapid pace of change in this arena.[88] The exponential growth and convergence of communications technology coupled with competition in standards setting by regional bodies and private sector software developers have served, in part, as the impetus for reform of ITU-T, which, until 2001, required that all standards be approved by Member states before they became official, even though most standards were developed in study groups comprised of the private sector.[89]

To help ensure that the ITU's preeminent role in this realm was not usurped, the Alternative Approval Process (AAP) was then adopted, which enables standards to be approved by the members of the study group that developed them, which is in essence the private sector. Adoption of the AAP has reduced the average time-to-market of ITU-T's recommendations by six months, for the traditional approval process averaged about 36.2 weeks, while the AAP is concluded in about 9.4 weeks.[90] Some, however, contend that this process is insufficient to curb the ascendance of a diverse range of private sector-led standard bodies and the ITU may become a mere rubber stamp for the outcomes of deliberations within the Global Standards Cooperation Group.[91] Regardless of the ITU's ability to retain its historical dominance in the standardization process, it must be noted that the AAP, which is now employed in virtually all Standardization Sector study groups, has enabled the ITU-T to effectively evade the sovereignty principle, which is arguably no longer germane in the current global

88 Mark W. Zacher and Brent A. Sutton, *Governing Global Networks: International regimes for transportation and communications* (Cambridge, UK: Cambridge University Press, 1996): 133.

89 Donald MacLean, "The Quest for Inclusive Governance of Global ICTs: Lessons from the ITU in the Limits of National Sovereignty," *Information Technologies and International Development*, 1, no. 1 (Fall 2003): 1–18.

90 Benjamin Fisher, "Chairman's Report," *The Point Newsletter of the United States ITU Association*, no. 8 (September 2003): 1–3.

91 Zacher and Sutton, *Governing Global Networks*, 153.

governance environment, and since it is from this principle that the ITU institutional framework stems, it is in the very framework of the ITU, its legal foundations, organisational structures and formal decision-making procedures, that the fundamental problems facing the organisation lie.[92]

Separating technical and regulatory functions has long been a source of debate as one can contend that all decisions have some public policy implications, so drawing the line between those issues in which the industry should have a major voice and those left to the regulators is a constant source of contention as is the role of the private sector. Indeed many consider the reforms instituted in 1992 inadequate, and, in regard to ITU-T, there were proposals, many of which originated in Europe, to further increase the role of the private sector by creating a separate forum, primarily governed and financed by private sector members.[93] The underlying issue confronting the ITU and the WTO as well as other international institutions is thus one of governance: who constitutes the decision-makers and to whom are these decision-makers accountable?[94]

## 3. *Regime Analysis*

### 3.1. *The Dynamics of Regime Change*

A regime is traditionally defined as a form of governance which comprises a set of norms, principles, rules and decision-making procedures that govern a particular issue area, such as the use of the global commons.[95] Space and the electromagnetic spectrum are part of our global commons, like seas, watercourses, and the atmosphere. The military and commercial utilization of space does not, however, generate widespread public interest or extensive involvement of non-governmental organisations as do other environmental issues. Radio frequencies or satellite orbital slots cannot be depleted as can fish stocks or minerals, but they are a relatively scarce and inordinately valuable resource that engenders competition for the best slots and specific frequencies. All wireless services use and thus require

92 MacLean, "The Quest for Inclusive Global Governance," 4.

93 Fisher, "Chairman's Report," 1–3.

94 Stiglitz, *Globalization and Its Discontents*, 18.

95 John Volger, *The Global Commons: A Regime Analysis* (New York: John Wiley & Sons, 1995): 18.

radio frequency spectrum. Demands for spectrum are escalating as a result of expanded applications in new technologies and as a result of the procompetitive, liberalised market mandated by agreements of the WTO.[96] The allocation, assignment and coordination of the radio frequency spectrum and orbital locations has grown increasingly complex, since having access to and authority to use the spectrum and specific orbital slots has implications for national security and foreign policy as well as economic, technological and societal implications for virtually every country and corporation in the world.[97] It is thus in this context of space and spectrum as part of our global commons that the satellite organisations under study can be defined as a common property resource regimes.

While this work does not intend to delve into the theories of international regimes, it is nonetheless necessary to differentiate meritorious, but contrasting points of view. Neo-liberals contend that interdependencies and mutual interests conspire to promote increased international cooperation. Certainly establishing a common property status for international space and airspace and creating international technical standards reduces the transaction costs of conducting commerce. This congruence of interest does not, however, devalue the validity of neorealist contentions that mutual interests are not sufficient for regime development. Neo-realist proponents contend that the presence of a dominant state or group of states which has the power to impose acceptance of the regime and compliance on other states is essential given that the hegemonic state or grouping will only support an international regime if it acquires greater relative gains.[98]

Cowhey contends that domestic politics and policies are the primary source of regime change as hegemonic players influence the change of institutional structures, affect the centralization of power in a regime, and alter the jurisdiction of the regime *vis-à-vis* that of an individual state.[99] Liberalisation of telecommunications as a domestic policy led, at the international level, to the introduction of trade in services into the General Agreement for Tariff and Trade (GATT) Uruguay Round 1986–1994. The series of accords that were approved in April 1994 included the General

96 Haller and Sakazaki, "Commercial Space and United States National Security," 73.

97 Ibid.

98 Zacher and Sutton, *Governing Global Networks*, 3.

99 Peter F. Cowhey, "The International Telecommunications Regime: The Political Roots of Regimes for High Technology," *International Organisation* 44, no. 2 (1990): 173–174.

Agreement in Trade in Services and a Telecommunications Annex. Adherents to these Agreements went on to establish within the institutional body of the WTO, which effectively succeeded GATT, an Agreement on Basic Telecommunications Services[100] and a Reference Paper, which placed national regulations in the multilateral arena for the first time. Arguably, more significant telecommunications deliberations now occur in the liberalisation-oriented WTO than in the ITU.

The privatisation of INTELSAT occurred in accord with WTO policies as well as the domestic policies of its dominant shareholding Signatories, most markedly the United States and its legislation, the Open-market Reorganisation for the Betterment of International Telecommunications Act or ORBIT Act, thus providing additional evidence corroborating Cowhey's assertions regarding institutional change. Although the ORBIT Act was initially proposed in 1997 in concurrence with the WTO General Agreement in Trade in Services, the legislation was agreed to or passed by both the US House and Senate in the 106th Congress on 24 January 2000. The ORBIT Act and subsequent privatisation of INTELSAT also supports Lyall's views regarding the role of the United States when he wrote some two decades ago, "the law of England as applied to maritime matters through the Admiralty Court had significant impact and became for most purposes international maritime law ... History, therefore, leads me to suspect that U.S. Law may become international space law ... By reason of its technical skills, its domestic market and its entrepreneurial attitudes, the USA is a major leader in space matters. That 'lead' may result in much U.S. Law becoming the language in which problems are discussed and solved."[101]

The existence of a hegemonic leader is not, however, as Volger states, the distinguishing characteristic of an effective regime, but the perception of high levels of mutual vulnerability interdependence.[102] In this regard, it was not considered viable for the United States, for example, to have left INTELSAT, as it had UNESCO, due to economic dependence on the vital, global satellite system. Cognizance of this position may lead to compromise, however, equally arguable is that it portends for hegemonic states,

100 Agreement on Telecommunications Services, *done* Feb. 15, 1997, ATS 1998 No. 9, 36 I.L.M. 354, entered into force 5 Feb. 1998.

101 Francis Lyall, *Law and Space Telecommunication* (Aldershot: Dartmouth Publishing Group, 1989): 419.

102 Volger, *The Global Commons*, 212.

notably the United States, to strive for greater influence over other regime members. The interests and activities of states, particularly of dominant states, and corporations are often indistinguishable, and Cox views the senior officials of such institutions as forming a transnational managerial class who have created a globally hegemonic business culture whose interests transcend the state-market dichotomy, and which affect changes within a given international institution.[103]

Issues of stability and change have long been of concern to regime theorists, and in this context it should be noted, for example, that the re-positioning of ITSO as overseer of Intelsat, Ltd.'s public service and lifetime connectivity obligations, arguably, signifies the adaptive process of the regime; but one can, however, contend that the privatisation of the satellite system, INTELSAT, represents actual regime change as it entails an alteration of norms and principles. While reform of ITSO did not include changing the achievement of consensus, the privatisation of the satellite system entailed an utter alteration of its structure and purpose as it seeks to provide end-to-end services through its established international network of teleports, leased fibre and points of presence (PoPs) for corporate business communications as well as for the broadcasting and entertainment industries. In such lucrative and competitive markets, it becomes necessary for Intelsat, Ltd. to give priority to its key customers, such as major Internet service providers, cable network, and international record carriers, rather than rural networks in developing countries, though such service provision, while less commercially viable, is of vital importance.[104]

### 3.2. *Satellites and Developing States*

Communication satellite systems, since their initiation in the late 1950s, have required extensive international oversight, especially transborder coordination of spectrum assignments and operations. Global satellite services were provided almost exclusively via international consortia, notably INTELSAT, throughout the first few decades of this industry, that is, until the early 1980s, when privately owned and operated systems were authorised.[105] Commercial applications, employing new generations of satellite

[103] Cox, *Production, Power and World Order*, 359–360.

[104] Daya Kishan Thussu "Lost in Space," *Foreign Policy*, 124 (May/June, 2001): 71.

[105] Schiller, *Digital Capitalism*, 66–67.

technology, steadily moved apace such that by 1998, the year in which privatisation of INTELSAT and INMARSAT moved apace, there were some 530 various types of satellites and 180 commercial communication satellites in geosynchronous orbit.[106]

Despite the growth in geostationary applications, the "world stood on the threshold of an unparalleled boom in satellite system and service development. US satellite makers planned to build and launch 1,700 satellites (mostly low-and medium-earth orbit models) over the decade beginning in 1998, worth a projected $121 billion – a rate of investment that would approach the level of US federal highway spending. A handful of mainly US-led consortia, headed by aerospace manufacturers, planned to ring the planet with next-generation multi-satellite systems."[107]

Indeed, the satellite industry witnessed continual growth throughout the 1990s, and, in response to competitive pressures in the satellite marketplace, the privatisation of intergovernmental satellite organisations.[108] Until the late 1990s, the majority of satellite systems providing non-military applications, namely, INTELSAT, INMARSAT, and EUTELSAT, continued to be owned and operated by multilateral treaty organisations. Each of these satellite systems had been comprised of government entities representing the nations that were party to the treaty and commercial companies that operated and used the services of the respective satellite systems.[109] The neo-liberal precepts of privatisation and liberalisation have, however, substantially altered the commercial communications satellite industry, drawing it in two related directions: toward multinational alliances and globalisation, as operators seek to secure requisite licenses and entry into previously restricted markets and expand in existing ones.[110] Privatisation has also allowed new entrants into the industry as private equity investors replaced sovereign governments as owners. It also placed the former intergovernmental satellite systems under a specific national jurisdiction.

For the developing world, the effects of this regime change were quite dramatic. Given that the telecommunications network in most developing countries does not sufficiently extend to rural areas and links that presently

106 Ibid., 67.

107 Schiller, *Digital Capitalism*, 67.

108 Haller and Sakazaki, "Commercial Space and United States National Security," 48.

109 Ibid., 48–49.

110 Ibid., 51.

exist are used to capacity, resulting in traffic congestion, the states can ill afford the capital investment required to extend these lines. In spite of the declining cost of fibre, it is not yet economically feasible in thin switched access routes. Satellite systems, however, make it possible to provide point-to-multipoint or point-to-point communication services to any site with cost independent of distance. Satellites are completely cost-insensitive to distance anywhere within the coverage area of a specific satellite, whether the two ends of a link are separated by ten miles or 3000 miles, the cost is the same. For virtually every other form of communications, distance is one of the key elements in providing a telecommunications link, and an important consideration for rural communications, thus many developing countries rely upon satellites for switched interconnections.

Furthermore, installation and maintenance of the ground segment are required only at the terminal site rather than at multiple points in between. Since any other point in the network is reached in a single 'hop' through the satellite, reliable interconnection depends solely on the two stations involved. This contrasts with terrestrial systems which depend on a series of sequential links, the failure of any one of which can disrupt the connection. Failure of any one earth station or dish affects only that location and has no negative impact on the rest of the network; it remains intact, thus reliability and access to remote areas are facilitated. This makes the satellite system a most robust system and one independent of trunking hierarchies. Facilities can be placed in those areas most needing communication services, regardless of distance from the urban centre, which reduces the bias of conventional terrestrial systems. Satellites can also enable immediate connectivity in short-term, emergency situations and can be employed with minimally developed infrastructure, such as poor roads and scarcity of prime power, and are not affected by rugged or inhospitable terrain, such as mountains, jungles, deserts, ice and snow.

Satellites are thus a critical communications component for many developing countries seeking to strengthen and diversify their economies, since the telecommunications sector is decisive in enabling countries to achieve social and economic development goals as well as compete in the international economy, as effective use of electronic communication permits improved coordination in the distribution and production of goods and services. Both case studies and statistical analyses have indicated a strong correlation between telephone penetration rates and GNP per capita. Since markets depend on the timely flow of information, the telecommunications network is arguably the most basic form of infrastructure, with a pervasive effect on the efficiency of an economy. The telecommunications infrastructure serves as a platform and a catalyst for other industries, thus the sector facilitates forward and backward linkages within the economy. It is an essential factor, a requisite for most businesses, especially foreign firms considering direct investment in a developing state. Since satellites can establish immediate multipoint connectivity without investments in physical conduit, they well serve governments, corporations, non-governmental organisations and others operating in those developing countries that lack an established domestic terrestrial system.[111] Countless numbers of people around the world in developing countries, however, remain without access to telecommunications services. Although the ITU recognises "the right to communicate" as fundamental, this right has proven challenging to ensure, especially as the world is witnessing the transformation of values in conjunction with the larger, subtle, but fundamental shift of power from nation states to transnational corporations.

### 3.3. *The 1990s – Witness to Continual Regime Change Via Neo-Liberalism*

The needs of transnational corporations for global economic restructuring have largely served as the impetus for rapid technological change and the embrace of the neo-liberal principles of privatisation and deregulation, which permeated the decade of the 1990s.[112] The United States observed the Clinton administration's passage of the Telecommunications Act of 1996 and subsequent mega media mergers. Although the legislation was passed in the name of competition, consolidation in the telecommunications and satellite industries is generally what transpired. The decade witnessed "a growing concentration of corporate investment and modern conservative tendency in the various apparatuses of public opinion formation, in the mass media in particular, and in the modern means (for example public relations and political advertising) for mobilizing a popular consensus regarding the competing value orientations of 'the market' versus 'state welfare' in the allocation of social benefits."[113] As media corporations grew increasingly vertically and horizontally integrated, controlling more "public" airwaves or outlets, they disseminated these neo-liberal

---

[111] Michael Einhorn, "Restructuring and competition in international telecommunications: the case of INTELSAT," *Information Economics and Policy* 10, no. 2 (1998): 201.

[112] Stolfi and Sussman, "Telecommunications and Transnationalism," 65.

[113] Ibid., 51.

messages through, what Umberto Eco labels, the language of "eternal fascism," whose purpose is to produce an "impoverished vocabulary" and an elementary syntax so as to limit the ability of the populace for complex and critical reasoning.[114] In manufacturing consent, the dominant media has thus not only failed to perform an essential public service, a service vital to a fully functioning democracy, it has deteriorated into a combination of commercialism, propaganda, and entertainment, and public discourse has been debased as it shifted from concerns of equity to efficiency.[115] This degradation of the media and the larger public sphere is due to neo-liberalisation as it required, both politically and economically, the construction of a neo-liberal market-based populist culture so as to ensure the palatability of wholesale privatisation.[116]

If privatisation was in vogue in the 1980s, by the 1990s it was status quo. The decade of the 1990s witnessed unprecedented levels of privatisation, with international capitalists as the primary beneficiaries. British Telecom had suggested in the early 1990s that INTELSAT be fully privatised, which was followed by a similar proposal by the U.S. Communications Satellite Corporation (COMSAT) in 1993. By 1998, the rate of global privatisation was roughly doubling each year.[117] This "privatisation revolution" coincided with accelerated globalisation, which, within the framework of neo-liberalism, represented the triumph of the economy over politics and culture and the hegemony of capital over all social domains of life, equating democracy and freedom with the unrestricted ability of markets to govern economic relations with little to no government regulation or oversight. Within the West's conceptual universe, the notion of democracy came to be closely, though faultily, linked to that of privatisation.[118] As neo-liberalism dismantled the historically guaranteed social provisions provided by the state, the market became the *de facto* international model for efficient government and business practice with the result being that private companies, often foreign ones, became the purveyors of public services as the state curtailed its role in meeting social needs.[119] The state is essentially in

114 Henry A. Giroux, *Against the Terror of Neoliberalism: Politics Beyond the Age of Greed* (Boulder and London: Paradigm Publishers, 2008): 27.

115 Ibid., 25.

116 Harvey, *A Brief History of Neoliberalism* , 42.

117 Singer, *Corporate Warriors*, 67.

118 Ibid., 67.

119 Ibid., 66–67.

the process of abandoning its commanding heights, and all of its most characteristic institutions are in decline—social security, justice, education, the military, and state owned enterprises, including telecommunications.[120] Modelling Thatcher's privatisation of British Telecommunications in 1984, the 1990s ushered in the privatisation of the telecommunications sector in numerous countries as well as that of the international satellite organisations. Indeed, there has been a tremendous increase in multinational alliances and globalisation across the entire commercial space sector, especially in the communications satellite segment.[121]

## 4. *Conclusions*

As we witness a redefinition of the public sector and the commons that is reflective of the tenets of neo-liberalism and as globalisation increases the interdependence among the people on this planet, so is enhanced the need for transparent international institutions to oversee global public goods and services, including the global commons.[122] Since the orientation of an institution is inextricably linked to those to whom it is responsible, voting rights are of paramount concern as voting determines whose voices are heard.[123] It is thus imperative that developing states ensure their voting rights in institutions of global governance, since the dominant ideology of neo-liberalism associates any idea of self-reliance as a symptom of regressive protectionism.[124]

There is, however, no democratic influence, let alone accountability, at international institutions such as the IMF, the WTO, and the World Bank.[125] Although Amin contends that respect for national sovereignty must remain the cornerstone of international law and that the locus of these laws should be the United Nations,[126] this principle is arguably undermined by the entry of private corporate actors and consequently obsolete. Indeed, states are witnessing a weakening of their autonomy as they gradually accept constraints on internal political control as a result of participation in

120 Ibid., 68–69.

121 Haller and Sakazaki, "Commercial Space and United States National Security," 58.

122 Stiglitz, *Globalization and Its Discontents*, 224.

123 Ibid., 225.

124 Amin, *Obsolescent Capitalism*, 116.

125 Harvey, *A Brief History of Neoliberalism* , 205.

126 Ibid., 116.

international regimes whose central purpose, increasingly determined by transnational firms, is to capture economic gains.[127]

Organisational reforms of international institutions are hence required to encompass not only corporate concerns, but to provide an equal voice and right to representatives of civil society. Alternatives to the demise of intergovernmental organisations and the ascent of corporate power are limited, however, by the authority structures underpinning the United Nations system, whose global governance, as evident in the WTO, has been complicit with the dominance of neo-liberalism as the rules and laws of its legal regime, formulated by the United States and other hegemonic industrial countries, facilitate the reproduction of capitalist social relations and thus the continued privatisation of the global commons.[128]

Just as a privatised commons is no longer a commons, the privatisation of the intergovernmental satellite organisations and their transformation into private equity acquisitions removes them from broader societal concerns. The pro-competitive privatisation process of INTELSAT and INMARSAT, in accord with the ORBIT Act, including their sales to private equity firms, was largely undertaken to benefit US corporations, not to create a satellite system better able to meet the telecommunications needs of the economically underprivileged countries of the South, to whom the companies bear little to no responsibility. As private equity acquisitions, the satellite operators can claim no direct accountability to citizens. They are, though, committed to the inclusion of the corporate class as their sole accountability is to shareholders.

While one can contend that commercial imperatives override access to the global commons as consolidation in the satellite industry has resulted in fewer large constellations of fleets providing services, the most salient change produced by neo-liberalism is in the international regimes themselves, be it their pseudo-privatisation, that is, the predominant influence extended by the private sector over institutions of global governance, such as the ITU and WTO, or their outright privatisation, as in the case of the former intergovernmental treaty-based satellite organisations. In sum, it is difficult to overstate the breath of neo-liberalism since neo-liberal theory has become hegemonic.

127 Zacher and Sutton, *Governing Global Networks*, 227.

128 Soederberg, *Global Governance in Question*, 14 and 46.

# Chapter Two

## Inmarsat: In the Forefront of Mobile Satellite Communications

*David Sagar and Patricia K. McCormick*

### 1. *Introduction*

The International Maritime Satellite Organisation (later renamed the International Mobile Satellite Organisation) (INMARSAT) was established in the late 1970s on the initiative of the International Maritime Organisation (IMO) (formerly the Intergovernmental Maritime Consultative Committee (IMCO)), a United Nations specialized Agency, which had sought to utilize space technology to enhance worldwide maritime communications and safety of life at sea.[1] Utilization of space at that time was considered to be primarily a governmental, rather than a private sector responsibility, and thus INMARSAT was formed as an intergovernmental organisation, albeit one with commercial aims, financed and managed by national telecommunications authorities. Once established in 1979, the organisation proved an immense boon to the maritime community, and throughout the 1980s its services and its revenues grew substantially, and its mandate was extended to provide both aeronautical and land mobile satellite communications.

In 1989, INMARSAT held a tenth anniversary international conference and space technology exhibition in London at which its achievements were celebrated, but the encroaching decade was to usher in change. During the 1990s, the private sector involvement in space exploitation, growth of competition, and the need for a more dynamic commercial

---

[1] Please note that henceforth in this document, INMARSAT shall refer to the satellite organisation prior to privatisation and Inmarsat will refer to the private operator and its subsidiaries after its privatisation. See also Section 9.6. *infra*.

approach led to INMARSAT's transformation into a private national corporate entity as provider of its services, with an intergovernmental organisation acting as a regulator of certain services deemed vital to the international public interest.

The process took some nine years, and it was quite unlike the familiar privatisation of national telecommunications entities. As a global intergovernmental organisation (IGO) serving public interests, the transformation of INMARSAT into a private sector corporation, with an international regulator, was fairly unique in international corporate life. It reflected the dismantling of telecommunications monopolies generally, and similar changes were undertaken by INTELSAT and EUTELSAT.

For three decades, INMARSAT and its successor organisation has served the mariner, the aviator and the land based user with high quality communications, using modern mobile satellite telecommunications technology. It has contributed to safety of life, media communications, peacekeeping and disaster relief worldwide. In the 21st century, the Inmarsat enterprise has proven successful in harnessing the constantly expanding satellite telecommunications technology for many kinds of mobile users, for commercial, social and humanitarian purposes. This chapter outlines the background and special features of the INMARSAT organisation, the particular forces which led inexorably to its transformation, the key elements of the new structure, the differing interests of the membership and the main political, commercial and legal issues encountered.

## 2. *Establishment and Purposes of Inmarsat*

An International Conference was convened by IMO in London in 1975–76 to determine how maritime satellite services could be organised on a global basis. It adopted INMARSAT's constituent instruments, namely, a Convention[2] and Operating Agreement[3] for the establishment of the International Maritime Satellite Organisation (INMARSAT) as an international, intergovernmental organisation. Both instruments entered into

---

[2] Convention on the International Maritime Satellite Organisation (INMARSAT)(hereinafter INMARSAT Convention), entered into force 16 July 1979, 31 U.S.T. 1, 1143 U.N.T.S. 105.

[3] Operating Agreement on the International Maritime Satellite Organisation (INMARSAT) (hereinafter INMARSAT Operating Agreement), entered into force 16 July 1979, 31 U.S.T. 1, 1143 U.N.T.S. 213.

force on 16 July 1979.[4] Three earlier proposals had been to establish an international maritime communications satellite consortium, and to set up a "users group" within IMO, and to provide the services as a specialized service by INTELSAT, but these had been rejected for various legal, commercial and political reasons.

The INMARSAT Convention defined the original institutional purpose of INMARSAT as providing the space segment necessary for improving maritime communications, thereby alleviating distress and improving safety of life, and enhancing management of ships, public correspondence services, and radio-determination (navigation) capabilities. This competence was subsequently extended to the provision, as practicable, of aeronautical and land mobile communications.[5] Reflecting these service extensions, the name was changed in 1994 to the International Mobile Satellite Organisation (INMARSAT). Certain key principles and public service obligations of the organisation were laid down in the Convention, and these determined the direction of INMARSAT's policies, namely:

- providing global maritime distress and safety services;
- serving all areas of the world where there is a need for satellite communications;
- ensuring non- discrimination on the basis of nationality in providing services;
- acting exclusively for peaceful purposes;
- operating on a sound, commercial, cost recovery basis; and
- complying with the international regulatory requirements of IMO, the International Civil Aviation Organisation (ICAO) and the International

4 Stephen Doyle, "INMARSAT: The International Maritime Satellite Organisation – Origins and Structure," *Journal of Space Law* 5 (1977): 45–63; Nandasiri Jasentuliyana, "The Establishment of an International Maritime Satellite System," *Annals of Air and Space Law* II (1977): 323–364; H.H.M. Sondaal, "The Current Situation in the Field of Maritime Communication Satellites: INMARSAT," *Journal of Space Law* 8 (1980): 9–39; David Sagar, "INMARSAT," *Annals of Air and Space Law* VII (1982): 505–507.

5 David Sagar, "INMARSAT," *Annals of Air and Space Law*. X1 (1986): 331–333 and "INMARSAT," *Annals of Air and Space Law* XIV (1989): 473–475; Wolf Von Noorden, "Space Communications to Aircraft: A New Development in International Space Law," *Journal of Space Law* 15 (1987): Part I, 25–34 and Part II, 147–160; and Wolf Von Noorden and Phillip Dann "Land Mobile Satellite Communications: A Further Development in International Space Law," *Journal of Space Law* 17 (1989): Part I, 1–11 and Part II 103–113.

Telecommunication Union (ITU) relating to maritime and aeronautical safety services and the use of the frequency spectrum.[6]

## 3. *Institutional Structure*

INMARSAT was created as an intergovernmental organisation, operating on a commercial basis and having public and private participation. Like INTELSAT before it, and later EUTELSAT, it was a curious hybrid, combining the structure of a classic intergovernmental organisation performing public service functions with commercial international public telecommunications services. This reflected the fact that most telecommunications entities at the time were State-owned monopolies, and many Governments wanted to retain some control over the provision of its functions.

The membership of INMARSAT comprised Governments which were Parties to the INMARSAT Convention, together with their designated national public or private telecommunications operators, which were Signatories to the Operating Agreement. The organisation functioned effectively as a co-operative. The Signatories (rather than member governments) financed the system in proportion to their investment shares, which were determined on the basis of their country's individual utilization of the system. Signatories did not have limited liability.[7] The largest Signatories participated in the management of the system through the Council. This differed from a typical board of directors in size, and because voting was weighted according to investment shares.

INMARSAT had three organs: the Assembly, the Council, and the Directorate. The Assembly was composed of all Parties to the Convention. It was primarily concerned with matters affecting member States in their sovereign capacities, and with general policy and long term objectives of the organisation. It also ensured that INMARSAT's activities were consistent with the Convention, the United Nations Charter and other

---

[6] David Sagar, "INMARSAT," *Annals of Air and Space Law* VI (1981): 587–589; Wolf Von Noorden and Phillip Dann, "Public and Private Enterprise in Satellite Telecommunications: The Example of INMARSAT," *Proceedings of the Twenty-Ninth Colloquium on the Law of Outer Space* (an International Institute of Space Law Colloquium), Innsbruck, Austria (October 4–11, 1986): 193–197.

[7] Art. XI, INMARSAT Operating Agreement, and *supra*, n.2.

international agreements binding upon INMARSAT. It elected four representatives to the Council with the mandate of ensuring equitable geographical representation. It expressed views in the form of non-binding recommendations with respect to technical compatibility and potential economic harm from separate satellite systems. It also decided on proposals to amend the Convention. It normally met every two years, and each Party had one vote.

The Council consisted of eighteen Signatories, based on investment shares, and four Signatories elected by the Assembly to ensure broad geographical representation with due regard to the interests of the developing countries. Decisions were taken by consensus or by majority decision based on weighted voting according to investment shares. The Council was the management organ of the organisation, mandated to provide for the space segment (the satellites and related ground control facilities) necessary to enable the organisation to carry out its purposes in the most economic, effective and efficient manner. It also determined the procurement policies, criteria and procedures for approving land-based and mobile earth stations, financial policies, determination of Signatory investment shares, annual budgets, charges for use of the space segment, and consultative arrangements with user groups, in addition to other responsibilities.

The Directorate, which was based in London, was headed by a Director General, appointed by the Council, and was the chief administrative organ, with operational responsibilities for the organisation.

## 4. *The Inmarsat System*

The INMARSAT system, which comprises the space segment, earth stations on land, and mobile earth stations, has grown extensively over the years. Its space segment comprises geostationary satellites located over different regions of the world, along with supporting ground control facilities. INMARSAT leased its first generation of satellites, and has since procured second, third and fourth generation satellites. The early generations had global beams, each giving coverage over approximately one-third of the earth's surface. The third and fourth generations also have spot beams, covering smaller areas where communications traffic is dense.

The present network of some eleven satellites includes the Inmarsat-4 series, the most advanced commercial mobile communication spacecraft of their kind, flying in geosynchronous orbit 35,786 km in space. They give full global communications coverage, with the exception of the polar regions beyond 75 degrees latitude. The Inmarsat-4 series is expected to continue in commercial operation until about 2020.

The communications capacity of the satellites is provided to earth stations owned and operated by independent entities. These stations receive and transmit communications between the mobile users and land points, via the satellites. The mobile earth stations are operated by the users of the system on ships, off-shore platforms, aircraft and on land. Inmarsat charges the operators of the land earth stations for the price of the utilization of the space segment. The land earth station operators charge the mobile end-users for the full cost of the calls, including the cost of landline connections.

## 5. *Use of Inmarsat Services*

The Inmarsat system of today offers a range of communications services comprising direct-dial telephone, telex, facsimile, electronic mail, and broadband data services for commercial use and public correspondence; high quality audio, compressed video, and video conferencing; and distress and safety communications, including distress beacons. Governmental communications, news-gathering entities and emergency services often operating in remote areas, all benefit from the latest technologies employed by Inmarsat. The recent Inmarsat-4 satellites have supported the Inmarsat Broadband Global Area Network (BGAN) mobile data communications at up to 492 kbits/s for internet access, mobile multimedia and other advanced applications.

Extensive use is made by world shipping of the INMARSAT system for general radiocommunications and distress and safety communications. On 1 February 1982, INMARSAT started providing services to around 1,000 ships with ship earth stations (SESs) which had previously operated through the Marisat satellites of the US COMSAT Corporation. Most of the world's merchant ships, including many naval ships and small government, leisure and other craft are now equipped with INMARSAT SESs.

The improvement of distress and safety of life at sea communications has always been central to INMARSAT's mission. Inmarsat, like its predecessor, has an Agreement of Cooperation with IMO, pursuant to which it ensures that the INMARSAT system conforms to IMO's relevant standards, especially those relating to the requirements of the IMO's Global Maritime Distress and Safety System (GMDSS). GMDSS has transformed the existing maritime distress and safety system by relying substantially on automation and the use of INMARSAT satellite communications in addition to other services such as the Worldwide Navigational Warning Service and Search and Rescue Co-ordination Centres (RCCs). The INMARSAT system has enabled the mariner to communicate instantly with vessels anywhere in the world and to contact immediately a RCC to summon assistance, thereby making a major contribution to safety of life at sea and the efficiency of rescue operations. In addition to the maritime services to the world's shipping fleet, services are provided increasingly to a wide variety of land-mobile users, including those in remote and underdeveloped regions, and to an increasing number of aeronautical users.

## 6. *The Path to Privatisation*

Throughout the 1980s, the organisation prospered commercially. As the only global operator of mobile satellite communications, INMARSAT had an assured maritime market, and had, as noted, begun to provide for aeronautical and land mobile satellite communications. Revenues grew steadily, and the Signatories were content to invest the capital needed for new generations of satellites with little demur. However, from around 1990, questions were raised as to whether the institutional and business structures that had served INMARSAT well, could cope with the new trends in the telecommunications world. A number of its co-operative features and government involvement were ill-suited to the changing commercial privatised environment. Under Article 5 of its Convention and several provisions of its Operating Agreement, INMARSAT was required to operate on a sound economic and financial basis having regard to accepted commercial principles. In practice, this meant that the IGO operated on a cost recovery but not a profit-making basis. It provided space segment capacity on its satellites at cost to Signatories and other telecommunications entities which, in turn, competed with each other in offering services to end-users at tariffs fixed by themselves. Only the IGO itself was limited in its charging policy by the operational and financial principles set out in Articles 5 and 19 of the original Convention. The IGO was effectively the wholesaler of access to utilization of the space segment by Signatories and other telecommunications entities. Signatories and such entities were the retailers of services and could establish their own tariffs to the end users. In fact such tariffs generally encompassed (a) the IGOs charges for use of the space segment, (b) the cost of landline links charged by terrestrial communications operators, and (c) their own costs and profit elements. Signatories and other entities were thus in competition among themselves in setting end user tariffs.

INMARSAT, though having a separate international legal identity, did not confer limited liability on its Signatories, nor did it have many other normal corporate features, such as the power to raise capital for new ventures on world markets. In particular, factors that made changes to the INMARSAT structure inevitable included:[8]

- the advent of competing satellite systems which affected the risks associated with investment in the organisation;
- a need for a normal corporate structure, with limited liability for investors and a small, fiduciary Board able to raise capital from the financial markets for the major new ventures which would be able to stand up to the competition;
- increasing privatisation of the telecommunications industry generally;
- proposals by regulators including the European Commission for wider access to space segment capacity by service providers and users;[9] and
- demand by many members for removal of tax and other privileges and immunities to require the organisation to compete on a level playing field with other mobile satellite system operators.

INMARSAT members began to realise that the long-term financial viability of the organisation could be jeopardized if substantial institutional changes were not undertaken, and this led to the privatised entity which exists today. As noted, the privatisation process took some nine years to accomplish. It occurred in three broad phases, firstly, experiments with

8 Warren Grace, INMARSAT Director General, *Recent Developments in Mobile Satellite Services*, Speech to the INMARSAT Users Conference, London, England (May 12, 1997).

9 Under Articles XIV and XV of the INMARSAT Operating Agreement, applications for approval of earth stations, and for the utilization of the INMARSAT space segment, were required to be submitted to INMARSAT through or via Signatories, or to or via authorised telecommunications entities in non-member countries. Although demand-assigned services are Inmarsat's primary service configuration (i.e. access to the space segment on a call by call basis when required), access is also provided through leases for certain services, subject to availability of capacity.

internal changes to the IGO; secondly, a diversion from restructuring to consider whether INMARSAT should provide personal handheld satellite communications, an exercise which exposed the structural shortcomings and adversely affected a major commercial opportunity; and, thirdly, overcoming resistance to full scale privatisation in the face of competitive challenge.

### 6.1. *The First Phase – Initial Restructuring (1991–93)*

The first glimmering of change came at INMARSAT's Seventh Assembly Session in Lisbon in October-November 1989 when Olof Lundberg, INMARSAT's first Director General, spoke of the need for making structural changes to INMARSAT in order to free it to act in a more commercial manner, in response to the changing telecommunications environment.[10] The formal start of the restructuring process took place at INMARSAT's Eighth Assembly Session in Canberra in September 1991 when the Assembly set up an Intersessional Working Group (IWG) of Parties and Signatories "to review the objectives and processes of INMARSAT in view of the changing telecommunications environment and the challenges of competition."[11] During 1991–1993, the IWG concentrated initially on streamlining the provisions for granting limited protection to INMARSAT's maritime services and on reviewing other commercial, operational and organisational arrangements.[12] Complementary work was carried on by the INMARSAT Council and its working Groups on subjects of special concern to the Signatories.[13] The Council, in particular, reviewed in detail INMARSAT's financial structure including possible revision of the investment share system and the ability to raise capital for future space segment from external sources or from a limited number of Signatories on a voluntary basis.[14]

---

10 INMARSAT, Assembly/7/14, *Report of Seventh Session of the INMARSAT Assembly* (Oct. 31–Nov. 2, 1989): section 10.5 (b).

11 INMARSAT, Assembly/8/13, (Germany), and Assembly/8/14 (Australia), *Eighth Session of the INMARSAT Assembly,* (Sept. 24–26, 1991).

12 INMARSAT, Assembly/8/16, *Report of Eighth Session of the INMARSAT Assembly* (Sept. 24–26, 1991): section 8.4.3.

13 INMARSAT, Assembly/9/15, *Report of Ninth Session of the INMARSAT Assembly* (Oct. 5–8, 1993): sections 8.1.5 & 6, and INMARSAT, Assembly/9/1, *Report of the IWG to the Ninth Session of the Assembly* (Oct. 5–8, 1993).

14 Cf. Articles III(e) and V(e) of the Agreement Relating to the International Telecommunications Satellite Organisation (INTELSAT), Washington DC (12 Feb. 1973),

This phase witnessed studies, proposals and debate on a variety of topics including improving governance and decision-making; examining the economic impact on INMARSAT of performing its public service functions in a competitive age; improving access to the INMARSAT system; adapting to new markets, and reviewing pricing policies to meet competition. Steps were taken to improve INMARSAT's access to national markets where regulatory barriers existed; to improve confidentiality for INMARSAT's commercially sensitive information; to ease restrictions on INMARSAT's commercial use of its industrial and intellectual property rights; to consider INMARSAT's right to enter into joint ventures with private entities; to review the scope of privileges and immunities, and to simplify the amendment procedure in the Convention. The range of these subjects gives an idea of the wide-ranging search for a structure which would be suited to the new telecommunications environment yet would retain intergovernmental oversight.[15]

### 6.2. *The Second Phase – The Challenge of Satellite Personal Communications (1993–94)*

The second phase was something of a divergence from the mainstream restructuring effort, but the outcome reinforced the need for major changes.

---

and Articles III(f) and V(f) of the Convention on the European Telecommunications Satellite Organisation (EUTELSAT), Paris, France (1 Sept. 1985) which enabled those organisations to provide satellites separate from their own space segment, under certain conditions including financing of the satellites by those requesting them. It is not known if any such separate satellites have been provided.

15 INMARSAT, Council/43/30, *Report on the Future Shape of INMARSAT*, 43rd Session of the INMARSAT Council (July 16–22, 1992); INMARSAT, Assembly/9/1, *Report of the Intersessional Working Group (IWG) to the Ninth Session of the INMARSAT Assembly* (Oct. 1993); INMARSAT, IWG/3/11, *Comparison with a Commercial Company* (23–24 July 1992); INMARSAT, IWG/4/2, *INMARSAT Participation in Joint Ventures* (Nov. 12–14, 1992); INMARSAT, IWG/5/5, *INMARSAT's Privileges and Immunities* (Mar. 25–27, 1993). See also later studies by the Inmarsat Directorate on topics raised during the restructuring debate, namely, IWG/11/4, *Conflict of Interests,* dealing with the potential conflict of interests of Signatories in their different roles as investors, land earth station operators, Council members and participants in competing enterprises (July 11–13, 1995) and IWG/12/1, *Intellectual Property Rights,* dealing with the need to ease restrictions in the Convention on INMARSAT's use of its IPR (Sept. 26–28, 1995); as to INMARSAT's efforts to improve access to national markets, see David Sagar, "Mobile Satellite Communications: Challenging the Regulatory Barriers," *Journal of Space Law* 25, no. 2 (1997): 150–154.

As early as INMARSAT's Tenth Anniversary Conference in London in July 1989, the first Director General predicted that by the year 2000, a worldwide system of satellite personal communications (S-PCS) through handheld phones would be in common use, eventually in competition with INMARSAT.[16] By 1993, it had become apparent that INMARSAT's enviable position as sole global provider of mobile satellite communications could not last. Its financial viability and ability to maintain its public maritime distress services obligations would be threatened if it did not adapt to the new technology of satellite telephony to satisfy user demands. Competition loomed from other systems, in particular Motorola's Iridium system which was planning to offer telephone services worldwide to personal handheld phones through a constellation of some 77 satellites in low earth orbit (LEO).

INMARSAT itself had, around this time, started development on a range of S-PCS services, originally called INMARSAT-P (for personal) and later Project-21 (for the 21st century).[17] Institutional problems, however, arose when some INMARSAT Signatories refused to contribute to the large capital needed for the project, on a mandatory basis according to their investment shares as required under the Operating Agreement. They wanted to choose the level of their own investment. Some felt that the business risks associated with the undertaking were too great, particularly in the absence of limited liability for the Signatories, though access to external sources of capital was not possible under the Operating Agreement. Other Signatories, which were mainly maritime communications providers, saw S-PCS as primarily a land mobile service, and not of direct interest to them. Some Signatories or their Parties required a "level playing field" among all S-PCS competitors, and did not want INMARSAT-P to benefit from INMARSAT's privileges and immunities or by cross-subsidization from its other revenues.

Another concern was that the cumbersome governance procedure through the INMARSAT Council would be inadequate to manage the

16 Olof Lundberg, "Cutting the Cord," in "*Never Beyond Reach: World of Mobile Satellite Communications,*" ed. B Gallagher (London: INMARSAT Publications, 1989): 58–79.

17 Alan Auckenthaler, "Recent Developments at INMARSAT," *Proceedings of the Thirty-Eighth Colloquium on the Law of Outer Space* (Oslo, Norway: International Institute of Space Law colloquium, Oct. 2–6, 1995): 149, and David Sagar, "Recent Developments at INMARSAT," *Annals of Air and Space Law* XX1.II (1996): 421–423 and David Sagar, "Recent Developments at INMARSAT," *Annals of Air and Space Law* XXII (1997): 363–365.

project with the decisiveness needed to match competition. As a result, the INMARSAT Council decided that the INMARSAT-P service should be provided through a separate, affiliated, private limited company under national law, to be called ICO Global Communications (ICO). Many of its shareholders would be INMARSAT Signatories. INMARSAT would have a minority shareholding, and arrangements would be made for INMARSAT to act as a wholesaler of non-handheld maritime and aeronautical services, using ICO satellites.[18]

This decision raised the issue whether, under treaty law, an IGO could, in the absence of explicit authority under its Convention, create a national law affiliate and transfer business assets (that is, the INMARSAT-P and intellectual property) to it. The matter was resolved at the INMARSAT Assembly's Tenth (Extraordinary) Session in December 1994. The Assembly decided that the Council's action was consistent with the Convention. It interpreted the Convention in a dynamic and evolutionary way, similar to that of a national constitution, taking into account the development of S-PCS satellite technology and the competitive business environment as demonstrated by the emergence of competitors like Iridium, Globalstar and Odyssey.[19]

Notwithstanding the links with INMARSAT, ICO would have been able to compete for some of the same mobile satellite communications markets that INMARSAT served, and INMARSAT would have appeared to have lost a valuable business opportunity in not retaining the INMARSAT-P system within its service portfolio. Eventually, however, ICO failed, along with other LEO competitors, Iridium and Globalstar, due to insufficient market demand and certain operational shortcomings, rather than competition from INMARSAT. Another commercial opportunity was rejected in early 1996, when the INMARSAT Council decided not to undertake a global INMARSAT Satellite Navigation Service because many Signatories were not prepared to invest further capital in INMARSAT under the current institutional structure.[20]

---

18 INMARSAT, Assembly/10/7, *Report of the Council on the INMARSAT-P Affiliate*, Tenth Session of the INMARSAT Assembly (Dec. 5–9, 1994).

19 *Supra* note 16, and INMARSAT, Assembly/10/18, *Report of the Tenth (Extraordinary) Assembly Session* (Dec. 5–9, 1994): section 4.

20 INMARSAT, Council/59/SR/Final, *Summary Record of the 59th Session of the Council* (Mar. 26–29, 1996): section 16.1, and INMARSAT, Assembly/11/23, *Report of the 11th Assembly Session* (Feb. 27-Mar. 1, 1996): section 7.5.

### 6.3. *The Third Phase – The Final Lap (1995–1998)*

Once the INMARSAT-P issue had been disposed of, the attention of the membership turned back to restructuring, but the final decision on privatisation did not come easily with a variety of proposals, long negotiations and shifting positions arising. The factors which drove the need for structural change were summed up by Warren Grace, the second INMARSAT Director General, as increased flexibility for investment in new systems and programmes, greater speed of decision-making, and ensurance of equitable competition. The Council's task was to recommend a viable commercial structure. The role of the Intersessional Working Group was to ensure that the structure satisfied the many policy concerns of Parties. It was realised that investment in new systems would not take place without major structural change, and key elements driving restructuring were the threat to future financial viability from competition and the insistence on giving up privileges and immunities, together with the demand by Parties that the maritime safety services and other public obligations must continue under governmental oversight.

Several main restructuring models were considered:

- *a multi-tier structure*, involving a continuation of the existing organisation to provide current services, with a subsidiary or affiliate to provide new services; this was seen as likely to lead to another "ICO" solution:
- *a revitalized INMARSAT*, amending the IGO's governance and financing provisions;
- *a treaty-based international public corporation (IPC)*, which would retain the intergovernmental organisation, but replicate many of the features of a commercial corporate entity under national law. Weaknesses of this model were that limited liability for investors could not be guaranteed, and the prospect of a floating of shares on a stock exchange was not high; and
- *a national law company*, retaining the IGO to oversee the performance by the company of the public service obligations. This was the model finally accepted.

## 7. *The Restructuring Decisions*

The decisions on restructuring spanned three Assembly Sessions, specifically, the Eleventh Session (27 February-1 March 1996), the Twelfth Session

(20–24 April 1998) and the Thirteenth Extraordinary Session (23–25 September 1998). This section traces those decisions, followed by a summary in Section 8 of the special political, operational and legal issues which were confronted.[21]

The Eleventh Assembly Session agreed that there was an urgent need to change the organisation's structure to enable INMARSAT to remain commercially viable in the long term. The challenge it faced was how to reconcile fundamental restructuring of the IGO with preserving crucial governmental interests. The Assembly decided that five basic principles should underlie any new structure, namely:

- continued provision of services for the Global Maritime Distress and Safety System (GMDSS) co-ordinated by the International Maritime Organisation;
- non-discriminatory access to services;
- service to all geographical areas where there is a need, including rural and remote areas;
- peaceful purposes; and
- fair competition.

Certain essential elements would also need to be taken into consideration in any future structure, including preserving the intergovernmental character of the organisation; continuing Assembly oversight of the basic principles; ensuring broad ownership and investment; limited liability for shareholders; representation of developing countries, and removal of privileges and immunities.[22]

In February 1997, the United Kingdom Party formally proposed amendments to the Convention and the termination of the Operating Agreement. The amendments were designed to implement the national law company model referred to above, which appeared to have the widest support among

---

[21] David Sagar, "Recent Developments at INMARSAT," *Annals of Air and Space Law* XX1.II (1996): 421–423; David Sagar, "Recent Developments at INMARSAT," *Annals of Air and Space Law* XXII (1997): 363–365 and David Sagar, "The Privatisation of INMARSAT," *Proceedings of the International Institute of Space Law(IISL) 41st Colloquium on the Law of Outer Space* (Melbourne, Australia: International Institute of Space Law, Sept. 28 -Oct. 2, 1998). See also INMARSAT, Assembly/11/3, Interim Report of the IWG to the Assembly, Eleventh Session of the INMARSAT Assembly (Feb. 27-Mar. 1, 1996).

[22] INMARSAT, Assembly/11/23, *Report of the Eleventh Session of the Assembly* (Feb. 27-Mar. 1, 1996): section 8.

the membership.[23] The next stage of the process, as required by the Convention, was for the Council to express views to the Assembly on the Convention amendments[24] and to approve the termination of the Operating Agreement, subject to confirmation by the Assembly.[25]

The Twelfth Assembly Session approved the restructuring amendments to the Convention and confirmed the termination of the Operating Agreement. In brief, the INMARSAT system was to be operated by a privatised national law company, and its public service obligations would be subject to governmental oversight by an amended IGO.[26] The Thirteenth (Extraordinary) Assembly Session, 23–25 September 1998, finally decided that the amendments would be provisionally applied as from 1 April 1999, or such later date as the INMARSAT Council decided, pending and subject to entry into force of the amendments in accordance with the procedures set forth in the Convention and Operating Agreement. The Council subsequently confirmed the transition date of 15 April 1999.

## 8. *Principal Elements of the New Structure*

Twelve core elements comprised the new structure. These are as follows:

- The INMARSAT business would be conducted under a multi-corporate structure, consisting of a holding company and an operating company registered under English law, and would continue to be based in London;
- The former INMARSAT Signatories would receive ordinary shares in the holding Company in a cash-free exchange for their former investment shares. They would also have limited liability;
- All of INMARSAT's assets and commercial business were transferred to the operating Company;
- The Companies' objects were to continue to provide global, regional and domestic satellite services, especially maritime, aeronautical and land

23 INMARSAT, Note to INMARSAT Parties and Signatories dated Feb. 5, 1997.

24 INMARSAT, Article 34 (1) of the INMARSAT Convention and Article XVIII (1) of the INMARSAT Operating Agreement (1976).

25 INMARSAT, Council 71/SR/Final, *Summary Record of the 71st Council Session* (Mar. 10–13, 1998): section 8.

26 INMARSAT, Assembly/12/20, *Report of the 12th Assembly Session* (Apr. 20–24, 1998): section 8 and relevant Annexes.

mobile commercial services, and distress and safety and navigation services;

- The Companies would no longer have any privileges and immunities;
- The Companies would have the same status under national regulation and in IGOs such as the ITU and the World Trade Organisation (WTO), as any private competitor;
- There were some special features in the structure of the companies intended to reflect the varying interests of the former membership and the global scope of INMARSAT's operations: the fiduciary Board of Directors of the holding Company would have up to 15 members, including shareholder directors, independent directors, and three directors from smaller shareholders or developing countries; and there would be an identical Board in the operating Company;
- Other special features were that there would be a limit on shareholding by any one investor to 15 per cent of the issued capital, except that the former United States Signatory would be able to retain its existing share of about 22 per cent for the time being;
- There would also be a requirement for the company to hold regional meetings to consider local interests;
- During the initial 12 months after 15 April 1999, trading of shares would only be possible among the existing shareholders; thereafter, it would be possible for the Company to issue shares to multiple investors from any country and to strategic investors. The Company would make an initial public offering on appropriate stock exchanges within approximately two years after that date;
- The intergovernmental organisation (IGO) would continue to exist, with a radically amended Convention, but its only organs would be the Assembly and a small Secretariat. The purpose of this IGO would be to ensure that the Company continued to provide space segment for GMDSS and meet the other basic principles and public service obligations. The continuing existence of the IGO would be reviewed when there were alternative providers of GMDSS services; and
- The IGO would own a Special Share in the holding Company, entitling it to veto changes to specified parts of the Memorandum and Articles of Association that related to GMDSS and the other public service obligations.

There would also be a contract, called the Public Services Agreement (PSA), between the IGO and the Company, enabling the IGO to oversee the Company's performance of the basic principles and public service obligations, and, if necessary, take certain enforcement action.[27] The functions and evolution of the IGO after privatisation and its relationship with Inmarsat are further described in Section 10.

## 9. *Special Political, Operational and Legal Issues Arising During the Privatisation Process*

### 9.1. *Membership Policies on Privatisation*

The wide diversity in the large INMARSAT membership (86 States as at 15 April 1999) meant that the road to fundamental restructuring would not be easy. The diversity lay, *inter alia*, in the size of the member States and of their investment shares in INMARSAT; in the scope of their mobile telecommunications services (some predominantly maritime, others land mobile); whether their respective national telecommunications policies favoured a government monopoly or a regulated private service provider; and their differing perceptions of the role of IGOs as global operators.[28]

One category of members, namely developed countries with large investments at stake in INMARSAT and which had privatised their telecommunications industries, were committed to fundamental restructuring for the reasons referred to above. Another category comprised a mixed range of members, including developing countries with smaller investment shares. These countries accepted the need for varying degrees of change in governance and financial mechanisms, but favoured a continued operation of the system through the intergovernmental organisation which they considered would better protect their interests in the basic principles, including the maritime safety services and the needs of rural and remote areas. Some members were also concerned that a privatised, profit-oriented INMARSAT might decide that the markets in their countries were not worth serving, and could leave them without satellite services. Some members of this

---

27 INMARSAT, Assembly/13/INF/6, *Commercial Restructuring Documentation* and Assembly/13 *Report*, Annex 34, documents of the 13th Session of the INMARSAT Assembly (Sept. 23–25, 1995).

28 David Sagar, "The Privatisation of INMARSAT: Special Problems," *Proceedings of the 3rd ECSL Colloquium on International Organisations and Space Law* (Perugia, Italy: European Centre for Space Law in association with the European Space Agency, May 6–7, 1999): SP-442.

group were newer members and may have felt that their interests were better protected in an IGO, where they could participate equally with all members in the Assembly in deciding broad policy aims, in contrast to a Company where they would only have a small shareholding and little voice in governance.[29]

### 9.2. *The Land Earth Station Operator (LESO) Agreement*

One of the most crucial and difficult issues was to conclude a standard Land Earth Station Operator (LESO) Agreement, to be signed between the future Company and each Signatory LES Operator. INMARSAT was a wholesaler of space segment capacity to the LES Operators at cost, which provided the services to the end users and set the retail charges. As most LESs in the system were owned by Signatories who had also sat on the Council and took executive decisions on services and pricing, a certain conflict of interest was built into the pre-privatisation system. The LESO Agreement was a distributor agreement guaranteeing that for five years after restructuring, the Company would be committed to providing the existing range and type of services at a steadily reducing cost to those LES Operators. The Company would also, during that period, continue to act as wholesaler of such services and would not be able to compete as a LES operator or a reseller of services. The existing LES Operators sought the protection of the LESO Agreement in order to preserve their investments and business in their LESs, once their control of services and pricing policy in the INMARSAT Council had passed to the Company. For them, it was an essential element in the restructuring. Ideally, the Agreement should have reflected a balance of interests between the LES Operators and the Company. However, the Agreement could have inhibited the maximisation of the Company's business opportunities, for example by limiting flexibility in managing its distribution outlets and its service portfolio, and in constraining its charging policy. At least, in the long term, the Agreement did not restrain the introduction of new services by the Company, which is where its future viability lay.[30]

---

[29] Examples of the views of some individual Parties can be seen from papers or statements attached to INMARSAT, Assembly/11/13, *Interim Report of the IWG to the 11th Assembly Session* (Feb. 27-Mar. 1, 1996).

[30] INMARSAT, Assembly/13/INF/6, *Annex XI* (Sept. 23–25, 1998).

### 9.3. *Investor Value and an IPO*

A vital aim of restructuring for many Signatories, especially the large investors, was to maximise the value of their investment shares and to be able to trade in their shares freely, as well as to obtain access to external capital for current and future systems. Formerly, investment shares could not be traded, except in a limited way through annual redeterminations of shares, and if a Signatory withdrew from INMARSAT, reimbursement for its share would have been at book value. There was a strong demand that the Company should undertake to list its shares on a stock exchange within two years of restructuring. While this was debated at length and opposed by some Signatories, particularly smaller ones who wanted to delay dilution of their shareholding in the Company, the agreed result was to provide that the Company would work towards an Initial Public Offering (IPO) within two years of transition.

### 9.4. *Convergence with ICO Global Communications*

Concern was expressed that INMARSAT's links with ICO should not distort competition, for example, by a merger which would give them a dominant position in the S-PCS market or monopolize the use of the limited, valuable mobile-satellite frequency spectrum available for those services.

### 9.5. *Composition of the Company Board*

Various corporate issues were also prominent in the debate. A hard-fought compromise on the composition of the Board was reached, balancing the need for a workable size on the one hand, with independent directors and representation for small and developing countries, as described above.

### 9.6. *Use of the Name "INMARSAT"*

The view was expressed that the use of the name "INMARSAT" by the Company would confer a competitive advantage on it and cause confusion with the IGO. The Council, conscious of the global reputation of the name, established by Signatories at great cost, insisted that it was an integral part of the business passing to the Company. Eventually, the Parties decided that the legal ownership of the name would remain with the IGO but it was licensed without charge to the Companies. The name was also dropped from the full title of the IGO, which remained "The International Mobile Satellite Organisation," later with the acronym "IMSO."

## 9.7. *Competition Law*

Compliance of the restructuring with competition law was, of course, necessary, and was one of the basic principles. Notification of the process was given to the European Commission which gave its endorsement. This decision was not unexpected, considering that the Commission's policy towards the international satellite organisations had been to improve access to space segment and remove tax exemptions and other competitive advantages.

## 9.8. *Limited Liability of the New Companies*

Limited liability for investors emerged as an essential condition of restructuring, taking into account the large capital which would be needed for future systems and the associated commercial risks in an age of strong competition. Examples exist of international public corporations (IPC) which have been established by treaty, with many structural similarities to national law companies, and in which the treaty limits the liability of shareholders. Expert advice, however, cast doubt on whether limited liability under a treaty would be recognised under domestic law generally. Even if the law of a State where the IPC established its headquarters explicitly accorded limited liability, this might not necessarily be recognised in other States where the IPC operated. It was considered that the limited liability of a corporate body established under a treaty was not sufficiently acknowledged as a general principle of law so as to guarantee automatic recognition in jurisdictions which were either not Parties to the treaty or in which the treaty had not been incorporated into domestic law.[31] The Signatories, who would be the initial investors in the Company, were not prepared to risk actions in any such jurisdiction, so the IPC model was not pursued.

31 Maurice Mendelson, *Jurisdiction-Limited Liability, Legal Opinion of Professor Maurice Mendelson QC*, IWG/16/INF/5, issued at the 16th Session of the INMARSAT Intercessional Working Group (Sept. 23–26, 1996): 1–12.

## 9.9. *The Intergovernmental Element of the New Structure*

One of the essential elements for restructuring laid down at the Eleventh Assembly Session was that the intergovernmental character of the organisation had to be preserved. The question at hand was how to privatise the system while implementing this element. There was at no time any challenge to the continued provision of the maritime distress and safety services or to the observance of the other basic principles. Indeed, those requirements were included in the Company's Memoranda and Articles of Association and the PSA. Although a private sector company would henceforth own and operate the INMARSAT system, the continuation of the IGO and the arrangements made under the PSA, as well as through the IGO's Special Share in the Company, satisfactorily met the requirement as to the retention of the intergovernmental character of the organisation.

The PSA itself was an unusual instrument in providing a regulatory mechanism whereby a large number of States would, through an IGO, oversee and enforce the activities of a nationally-incorporated multinational company. The PSA was governed by English law, and gave the IGO effective enforcement remedies, including specific performance and injunctive relief, in the event of failure by the Company to observe most of its obligations. The PSA also required the Company to fund costs of the IGO's Secretariat to the extent of £300,000 per year initially plus a contingency fund of £100,000 for enforcement costs. The PSA could be terminated by the company when competing systems were established which were recognised by IMO as satisfying its requirements for maritime distress and safety satellite communications. The prominence given to IMO's role under this Agreement was another unique feature of the restructured Inmarsat.

## 9.10. *Provisional Application of the Amendments*

The legal issue *par excellence* which shadowed the whole restructuring process was the need for provisional application of the amendments. The entry into force of amendments to the Convention adopted by the Assembly normally took some years, due to the requirement of formal acceptance of the amendments by a two-thirds majority of the membership, representing two-thirds of the investment shares at the time of the adoption of the amendments. As this delay would have defeated the commercial purposes of restructuring, there was need to resort to the concept of provisional application.[32] Provisional application is well established in interstate treaty practice, as reflected in Article 25 of the 1969 Vienna Convention on the Law of Treaties,[33] but its use by decision of a supreme body of an IGO such as the Assembly was not extensive. Analysis of the problem by treaty experts led the Council and the IWG to recommend that the Assembly decide to implement the amendments provisionally, subject to eventual formal entry into force.[34]

Consultations with Parties over this issue disclosed that, while many accepted provisional application, others were precluded by their constitutions from doing so without special legislation, particularly as the Convention did not itself provide explicitly for provisional application. For these countries, practical solutions were proposed whereby a provisional application decision by the Assembly would be subject to the internal laws of each Party. This, it was argued, would enable them to participate in the restructuring, including the exchange of their Signatories' investment shares for ordinary shares in the Company, without being forced to oppose the decision.

It was important that the Assembly's decision should be taken by consensus, given the substantial nature of the amendments which affected the rights and interests of all Members of the organisation, and the need for certainty in implementing the complicated legal arrangements to give effect to the restructuring. In the event, the Assembly took the historic step of deciding on provisional application of the amendments without a vote.[35] Individual statements made by a number of Parties during the debate relevant to the provisional application decision were attached to the official Report of the Assembly, in accordance with the Assembly's Rules of Procedure.[36] INMARSAT, now Inmarsat, Ltd., was thus effectively

---

32 *Supra* note 16.

33 Art. 25, Vienna Convention on the Law of Treaties, *done* May 23, 1969, 1155 U.N.T.S. 331. Available: http://untreaty.un.org/ilc/texts/instruments/english/conventions/1_1_1969.pdf.

34 Shabtai Rosenne, *Legal Opinion of Professor Shabtai Rosenne*, issued at the 8th Session of the INMARSAT Intersessional Working Group, IWG/8/3, *Annex II* (June 13–17, 1994); and David Sagar, "Provisional Application in an International Organisation," *Journal of Space Law* 27, no.2 (1999): 99–116.

35 INMARSAT, Assembly/13/*Report of the 13th (Extraordinary) Session*, (Sept. 23–25, 1998): section 4.2 and relevant Annexes. See also Article 34 (2) of the INMARSAT Convention and Article XVIII (2) of the INMARSAT Operating Agreement.

36 INMARSAT, Assembly/12/14 (Apr. 20–24, 1998).

privatised on 15 April 1999, although the amendments only came into force formally on 31 July 2001, binding on all Parties, including those which had not yet accepted them. The decision to apply the concept of provisional application to amendments to the constituent instruments of an intergovernmental organisation for its restructuring, mainly as a privatised entity, represented a major contribution to inter-State practice in this field, and was followed by INTELSAT and EUTELSAT as the trend towards privatisation continued.[37]

### 9.11. *Relations with IMO*

Close cooperation between INMARSAT and IMO led to the INMARSAT system being specified in the Safety of Life at Sea Convention (SOLAS) as satisfying IMO's requirements for its GMDSS. The intergovernmental character of INMARSAT gave IMO the necessary assurances of continuity of the system consistent with IMO's standards. After privatisation, the IMO's Maritime Safety Committee (MSC) expressed satisfaction that the continued government oversight of the maritime services under the PSA and the commitments in the Company's constituent instruments were sufficient to meet IMO's concerns.[38]

### 9.12. *Relations with Other International Organisations*

INMARSAT, as an IGO, had long cooperated with the IMO, ICAO and ITU and certain regional IGOs under Agreements of Cooperation, with observer status at meetings, as well as industry and user entities. This greatly facilitated INMARSAT's work in establishing its global services, compliant with the regulatory standards of those organisations. The transition to private sector status required a change in these arrangements for the Company, though they will remain in place for the IGO. The IMO and ICAO, as specialized UN agencies, did not accord observer status to or conclude formal cooperation agreements with private companies. The

[37] For a detailed review of evolving international practice concerning decision-making by consensus in international organisations, see Henry Schermers and Niels Blokker, *International Institutional Law*, 4th revised ed. (Leiden, The Netherlands: Martinus Nijhoff Publishers, 2003): 809–811.

[38] INMARSAT, Assembly/12/14 (Apr. 20–24, 1998).

situation at the ITU is different as there the Company was accorded the status of a Recognised Private Operating Agency.

### 9.13. *Other Transitional Arrangements*

In addition to the normal detailed process of transferring the assets and business to a new Company, a major due diligence process was undertaken by an investment bank, tax advisers and lawyers to advise on the likely acceptability of the restructuring arrangements to future investors, and to ensure the financial, economic and legal soundness of the transaction. Extensive negotiations were conducted with the United Kingdom Government to bring the organisation and its staff within the United Kingdom corporate and personal tax regimes, to establish the status of the Company at the ITU, and to obtain licenses for operating a space telecommunications system, including those needed under the UK Outer Space Act of 1986,[39] which implements the United Kingdom's obligations under the outer space treaties.

### 9.14. *The New Corporate Structure*

On 15 April 1999, the whole of INMARSAT's assets, business, operational functions and staff were transferred to a UK-based corporate structure ("the Company"). The operating company initially had a fiduciary board of 14–15 directors (including representatives of smaller and developing countries). The former Signatories held ordinary shares in the holding company, in proportion to their former investment shares, and the holding company undertook to list its shares on a stock exchange within about two years. The mechanism for ensuring continuance of the public service obligations and their supervision is set forth in Section 10. The new corporate flexibility enabled the Inmarsat company to diversify its services and seek new avenues of business. It was also able to take its long-awaited decision on procuring its Inmarsat-4 generation of satellites to secure the future viability of the Company.[40] The first new satellite was successfully launched from Florida in early March 2005.

---

[39] Outer Space Act, July 18, 1986, 1986 Chapter 38; *National Space Legislation of the World*, Vol. I (2001): 293; *Space Law—Basic Legal Documents*, E.I; 36 *Zeitschrift für Luft- und Weltraumrecht* (1987): 12.

[40] For the long-term business implications of the privatisation for Inmarsat and its future commercial prospects, see keynote speech by Warren Grace, Inmarsat Director

## 9.15. *Requirement for an IPO*

One of the conditions of privatisation was that the Company would undertake a listing of its shares on a stock exchange through an IPO within two years. The purpose of the listing was to bring in external capital for necessary investment, dilute the existing shareholding structure, which was based upon previous Signatory participation, and establish a market value for existing shares. This condition was particularly required by the United States which had enacted the ORBIT Act (Open–market Reorganisation for the Betterment of International Telecommunications) of 2000,[41] which bore upon both INTELSAT and INMARSAT, despite the fact that INMARSAT privatised prior to the enactment of the Act. The Act reflected the opposition of the United States to the use of ISOs for projects which they considered were better done by the private sector, without an unfair competitive advantage. One provision of the ORBIT Act required that the privatised companies conduct an IPO of equity securities that would "substantially dilute" the aggregate ownership of the former Signatories of INTELSAT and INMARSAT.[42] These entities were either private telecommunications companies or governmental agencies of the countries party to the intergovernmental agreement that formally established INTELSAT and INMARSAT, and these entities became shareholders of the respective companies upon their privatisation. Under the Act, time limits were set for the IPOs of Inmarsat and Intelsat, in fact, Inmarsat was originally required to conduct an IPO no later than 1 October 2000; however, both Inmarsat and Intelsat, Ltd. obtained successive US Congressionally-approved extensions as to the date by which they needed to conduct an IPO, thus giving the companies the opportunity to explore other options.

Ultimately, an amendment to the ORBIT Act was signed into law on 25 October 2004, providing an alternative method by which Intelsat and Inmarsat could comply with the privatisation requirements of the Act, that is, the dilution of share ownership, by adding new subsections (F) and (G) to Section 621(5). Following Inmarsat's acquisition by private equity funds advised by Apax Partners and Permira in December 2003, in December

General, *The Future of Mobile Satellite Communications Services*, Speech to the INMARSAT Mobile-Satellite Communications Conference, Marbella, Spain (Oct. 1998).

41 Open-Market Reorganisation for the Betterment of International Telecommunications Act (hereinafter ORBIT Act), 47 U.S.C. § 761 (2000).

42 ORBIT Act, S. 376, 106th Cong., 2d Sess. 2000: 5.

2004, the US Federal Communications Commission (FCC) issued an Order granting applications filed by Intelsat and Zeus Holdings Limited (Zeus), a company formed by a consortium of private equity funds comprised of Apax Partners, Apollo Management, Madison Dearborn Partners and Permira, to transfer control of certain FCC authorisations from Intelsat to Zeus. Then Intelsat, again imitating Inmarsat's actions, filed a Petition for Declaratory Ruling and attached certification, pursuant to Section 621(5)(F) of the ORBIT Act, that upon consummation of the transaction with Zeus, the company would be in compliance with the said requirement as it would achieve substantial dilution of the aggregate amount of former signatory financial interest in Intelsat.[43] In April and June of 2005, the FCC determined that Intelsat and Inmarsat, both of whom became private equity acquisitions, arguably to ensure the financial well-being of their largest shareholders, were in compliance with the alternative certification process of the ORBIT Act, and could thus forgo the requirement for an IPO and the public listing of securities, and that Inmarsat was no longer subject to the provisions of Section 602 that prohibited it from providing additional services.[44]

Inmarsat embarked upon new service offerings when the Commission granted it market access to the United States in 2001. Indeed privatisation provided the company with the opportunity to develop new services for the US market, resulting in the expansion of service options and providers for customers in the United States as Inmarsat distributors were given access rights to the United States. An assessment of the effects or the impact of the privatisation of Inmarsat, as well as Intelsat, on the US industry, jobs, and the global market is required by Section 646 of the ORBIT Act to be reported by the Commission to the US Congress. In this regard, Inmarsat contends that its services promote economic growth and job creation in the United States, as its system is used in managing the sustainability of fisheries and its portable terminals are employed by US companies working in remote regions in the areas of energy, mining exploration, construction and journalism.[45] In sum, in the decade since Inmarsat was granted broad access to the US market, the company has developed into a dominant player, accounting for more than a quarter of the estimated

43 In the Matter of Intelsat, Ltd., 20 F.C.C.R. 8604 (2005).

44 FCC Report to Cong. As Required by the Orbit Act, 22 F.C.C.R. 11347 (2007): 6 & 8.

45 Ibid, 18.

US$ 750 million the US government spends annually on commercial satellite capacity.[46]

In this and other respects, Inmarsat sought to be in full compliance with the ORBIT Act. Furthermore, the company, though not required to do so, conducted an IPO on the London Stock Exchange. Ordinary shares of Inmarsat began trading under the symbol ISAT as of 22 June 2005. With the IPO, the value of Inmarsat shares rose nearly 16 percent.[47] The IPO gave the company's two principal shareholders, leveraged buyout companies Apax Partners and Permira, a quick return on their late-2003 purchase of Inmarsat, and provided them and other Inmarsat shareholders with liquidity.[48] It also diluted the remaining interests of former INMARSAT Signatories and foreign government entities that owned INMARSAT shares.

The parent company is now Inmarsat plc, a UK company and its wholly owned subsidiary, Inmarsat Global Limited, based in London, is the main operating company. Today Inmarsat's shares are traded on the London Stock Exchange with the aggregate ownership by foreign governments nominal. The parent company has three executive directors and four non-executive directors. The operating company now has a Board of only four directors, one of whom represents the interests of developing countries.

## 10. *Functions of the International Mobile Satellite Organisation (IMSO) and its Relations with Inmarsat*

### 10.1. *Supervision of Inmarsat's Public Service Obligations – The Public Services Agreement*

One of the central questions in the privatisation process was how INMARSAT Member States could ensure that the new company would continue to carry out the public service obligations which were entrenched in the INMARSAT Convention. The possibility of regulating the company under a licensing system or legislation by the headquarters' State was

---

46 "Profile Rebecca Cowen-Hirsch, President, Inmarsat Government Services," *Space News* 21, no. 18 (May 3, 2010): 22.

47 Ibid.

48 Peter de Selding, "Inmarsat IPO Aims to foster Growth of Partners, Services," *Space News* (2005). Available: http://www.space.com/spacenews/archieve05/Inmarsat_060605.html.

considered to be inappropriate for the intergovernmental control required by States. The solution adopted was to retain the existing IGO in a restructured form to act as a monitor and regulator of the private company's public service obligations or basic principles. These obligations would be secured under the PSA, a legally-binding bilateral agreement between the restructured IGO (IMSO) and the company.

The PSA is governed by English law, pursuant to which the Company agrees with IMSO to perform the public service obligations, specifically:

- continued provision of services for GMDSS set up by IMO (INMARSAT was then the only system recognised in IMO's SOLAS Convention as the provider of certain mandatory satellite safety communications services for ships);
- non-discriminatory access to services by users of all nations;
- service to all geographical areas where there is a need, including rural and remote areas;
- the obligation to act exclusively for peaceful purposes; and
- fair competition.

The PSA also contains other obligations by the Company, namely, to undertake an IPO within two years, as described in Section 9.15, and to take account of or comply with the regulatory standards of the IMO, ICAO and the ITU. If necessary, IMSO may take enforcement action in the English courts to ensure compliance by the Company with the above obligations.

The PSA may be terminated by agreement among the parties, but it is expected to continue until such time as IMO confirms that there are alternative satellite systems providing the GMDSS services. This arrangement is unusual in that IMO, though not a party to the PSA, is named as the entity that ultimately decides whether its primary purpose has been fulfilled. Other controls were also built into the Inmarsat restructuring instruments to strengthen the IGO's supervision of the public service obligations. IMSO was given a Special Share in the Company which enables it to veto any changes in the objects and powers of the Company which would be inconsistent with the Company's obligations under the PSA. Provisions were also included in the memoranda and articles of association of the Company relative to the performance of those obligations.

Under the PSA, a Public Services Committee comprising representatives of IMSO and the Company was established to ensure on-going consultation and cooperation between them on all relevant PSA issues. Provision was made for all reasonable information about Inmarsat's activities, including engineering and related advice, to be furnished to IMSO. Annual reports are also made to IMSO on the performance by the Company of its GMDSS obligations under the PSA. The costs of running the IMSO Directorate are paid by the Company. Responsibility for the day-to-day implementation of the PSA and the amended IMSO Convention, and overseeing the obligations of the Company, is in the hands of a Director General, elected by the IMSO Assembly, who heads a small Directorate based in London.

### 10.2. *Exercise of the Supervisory Powers of IMSO Since Privatisation*

Since privatisation, the IMSO Director General has cooperated closely with the Company and also with IMO, in particular its MSC, in connection with the implementation of the PSA. The IMSO Director has sent regular reports to IMSO on the Company's activities and the IMSO Assembly has acknowledged that the Company has complied with its obligations under the PSA. At its Seventeenth Session, held in October 2004, the IMSO Assembly noted that the Company had continued to provide a sufficient quality of service to meet its GMDSS obligations under the PSA. This had also been confirmed by the MSC. The Assembly noted that the Company had satisfied the other public service obligations and commitments contained in the PSA, and that following INMARSAT's change of ownership, the good relationship between the new owners of Inmarsat and IMSO had been maintained. The IMSO Director General informed the Assembly that, in his view, the rights of IMSO Member States in relation to the public services obligations were sufficiently secured under the new arrangements.

### 10.3. *Amendments to the IMSO Convention and the Future of the PSA*

Since the restructuring, significant developments have taken place with respect to IMSO's mandate under its Convention and the PSA. The main change relates to the performance of the GMDSS services. IMO's mandatory regulatory requirements for shipping under the GMDSS are set forth in IMO's SOLAS Convention. Hitherto, Inmarsat has been the only satellite operator recognised under SOLAS as meeting the technical requirements of the GMDSS. This gave Inmarsat an effective monopoly on the provision of maritime satellite communications services; however, it was recognised at IMO that these arrangements would eventually need to change. By Resolution A.888 (21), dated 25 November 1999, the IMO Assembly adopted general criteria for provision of GMDSS services, enabling operators other than Inmarsat, which satisfied the criteria, to provide such services to ships.

IMSO agreed, at IMO's request, to carry out the oversight of future providers of satellite services for GMDSS. For this purpose, amendments were needed to the IMSO Convention to enable it to act as the supervisor of each new provider of maritime mobile satellite communications services for the GMDSS. For this purpose, "Service Provider" is defined as meaning any entity providing GMDSS services through a mobile satellite communications system recognised by IMO. Amendments to the IMSO Convention have also been made to enable IMSO to oversee, in addition to GMDSS, long range tracking and identification of ships (LRIT), which is a new initiative of IMO in order to improve maritime security, safety and environmental protection.

### 10.4. *Amendments Relative to Peaceful Purposes*

The obligation to act exclusively for peaceful purposes was an important provision in the original INMARSAT Convention. It was consistent with the UN principles relative to the peaceful uses of outer space and the creation of INMARSAT as a commercial IGO carrying out public service obligations. From its inception, INMARSAT received requests for authorising the use of INMARSAT satellite terminals by naval auxiliary ships and also by UN peacekeeping agencies. In the absence of a definitive interpretation of the meaning of "peaceful purposes," the INMARSAT Directorate adopted a pragmatic policy, endorsed by its Parties, of authorising the use of such terminals upon receiving written assurances from the Governments concerned that they would be used for peaceful purposes. Later, after further legal analysis, use of terminals were permitted if the eventual purpose was a peaceful one, and this included use by UN authorised peacekeeping forces, use in cases of legitimate self-defence under Article 51 of the UN Charter, and also use under the Geneva humanitarian conventions and certain other treaties.

The policy guidelines followed by the INMARSAT Directorate were as follows:

- use of INMARSAT by armed forces (military use) not involved in armed conflict or any threat to or breach of the peace is permitted;
- use of INMARSAT by UN peacekeeping or peacemaking forces acting under the auspices of the UN Security Council is permitted, even if engaged in armed conflict to accomplish their mission;
- use of INMARSAT by UN peacekeeping or peacemaking forces not acting under the auspices of the UN Security Council involved in international or non-international armed conflict (civil war) is not permitted, except in case of legitimate individual or collective self-defence against armed attack within the limitations established by the UN Charter,[49] Article 51; the latter exclude preventive action and self-help involving armed force in the absence of armed attack;
- use of INMARSAT by armed forces engaged in armed conflict is permitted for Distress and Safety communications, and for communications relating to the protection of the wounded, sick, shipwrecked, prisoners of war and civilians pursuant to the Geneva Red Cross Conventions of 1949, and the Protocols Additional to the Geneva Conventions of 1977; also permitted are personal and private, non-tactical communications by members of the armed forces; however, use of INMARSAT for non-tactical governmental communications related to or in support of the war effort that do not originate from governments of, and are directed to, armed forces engaged in activities sanctioned by the UN Security Council, or self-defence pursuant to UN Charter Article 51, is not permitted.[50]

INMARSAT terminals have been used extensively by media, international agencies and the military in war zones such as the Balkans and in both Iraq wars. The Company sought to free itself from these restrictions because the privatised Intelsat and Eutelsat were not subject to them, and therefore gained an unfair competitive advantage over Inmarsat. Furthermore the restrictions were unenforceable because the Company was not entitled to monitor the content of satellite communications under the ITU regulations, and would not have been able to determine whether a terminal was actually being used for non-peaceful purposes.

---

49 Charter of the United Nations, *entered into force* 24 October 1945; 24 U.S.T. 2225, 892 U.N.T.S 119 (1945).

50 Wolf von Noorden, "Inmarsat Use by Armed Forces: a Question of Treaty Interpretation," *Journal of Space Law* 23 (1995): 1–17.

Accordingly, the IMSO Convention was amended by removing the Company's obligation to act exclusively for peaceful purposes. The principle survives as an obligation on IMSO itself, effectively an unenforceable provision, as IMSO itself does not provide any communications services, and no longer has authority to regulate the Company's communications in that respect.

### 10.5. *Services in All Areas of Need*

This principle has been preserved in the amendments to the extent that IMSO shall seek, through existing international and national mechanisms dealing with technical assistance, to assist service providers to ensure that all areas where there is a need are provided with mobile satellite services, giving due consideration to the rural and remote areas. This shifts the direct burden of carrying out the obligation to IMSO rather than the Company, as it had been under the restructuring agreements.

### 10.6. *Non-discrimination in the Provision of Services and Fair Competition*

These obligations have not been carried forward into the amended IMSO Convention. The Company is, of course, legally obliged to observe international and national competition laws. In the case of non-discrimination, the removal of the obligation means that the Company could in the future decide to charge higher tariffs in regions where traffic is lighter, and this was among the concerns of smaller, developing countries at the time of privatisation. This removal of obligations could arguably, however, assist Inmarsat in becoming a more profitable firm.

## 11. *New Services and Acquistions*

Inmarsat, which currently operates a fleet of 11 L-band communications satellites in geostationary orbit, reported 2011 revenues of US$ 1,408.5 billion,[51] which, coupled with a readily manageable debt load, has left itself well positioned for continued growth across all three of its service sectors: land-mobile, aeronautical and maritime. Inmarsat's strong economic

---

[51] Inmarsat plc, *Annual Report* (2011): 1.

performance has also been aided by a decline in its capital spending starting in 2009, in conjunction with its 13 per cent reduction in operating costs at that time.[52] In addition to launching the Broadband Global Area Network (BGAN) in 2005, which in 2009 alone had grown 33 per cent in terms of revenue, Inmarsat is seeking to better compete in the satellite telephone markets as well.[53]

### 11.1. *Services*

At the end of 2005 with the first two of three of its next-generation Inmarsat-4 satellites in commercial operation, Inmarsat launched its BGAN service, the world's first mobile communications service to provide both voice and broadband data simultaneously through a single, highly compact device on a global basis. The Inmarsat-4 series of satellites had tremendously increased capacity compared with the Inmarsat-3 spacecraft, though, like their predecessors, they were equipped with a single global beam that covered up to one-third of the Earth's surface, apart from the poles. Each Inmarsat-4 satellite also generated 19 wide spot-beams that provided continuous coverage across the same region for Inmarsat's existing high-end services. Furthermore, new to the Inmarsat-4 satellites, were an additional 228 narrow spot-beams, designed to form the backbone of Inmarsat's broadband services, including BGAN, which delivered Internet and intranet content and solutions, video-on-demand, videoconferencing, fax, email, phone and LAN access.

Essentially, BGAN provided a broadband mobile office in the most remote locations in minutes, enabling users to access data applications at speeds up to half a megabit while simultaneously making a phone call. Compact and lightweight, BGAN terminals, manufactured by Hughes and Thrane and Thrane, were designed for simplicity and ease of use, for journalists, government personnel, aid workers, military personnel, and other users of mobile satellite communications. After plugging the briefcase-size BGAN terminal into any laptop computer with a standard USB cable (or using a Bluetooth or Wi-Fi connection), mobile users of all types have

---

52 "Inmarsat plc report Full Year Results 2009," Inmarsat, accessed Apr. 12, 2013, www.inmarsat.com/corporate/media-centre/press-releases/inmarsat-plc-reports-full-year-results-2009.

53 Ibid.

immediate voice and data connectivity regardless of the state of the terrestrial network. Indeed, after a devastating earthquake struck Haiti in January 2010, a team from Inmarsat-sponsored aid agency Telecoms Sans Frontières (TSF) were among the first responders, using reallocated spot beams from the INMARSAT-4 satellites to meet high demand from aid agencies and the military.

Further, BGAN supported all major VPN products and encryption standards, whether commercial or military grade, and users would select guaranteed data rates on demand, with a choice of speeds to suit one's application requirements. In January 2009 Inmarsat's third and final Inmarsat-4 Americas satellite entered into commercial service and Inmarsat then proceeded to complete its repositioning process of the other two INMARSAT-4 satellites, resulting in full global coverage for its unique broadband suite of services, ushering in a new era which it calls 'Broadband for a Mobile PlanetTM.'[54] Also, in effect by January 2009, as part of its project known as One World, One Number, Inmarsat, under an agreement with the ITU, returned the four ocean region codes (-871, 872, 873 and 874) used for its services since operations began in 1982, and switched to a single dialling code, +870, for all of its services. Inmarsat also announced, in April 2009, BGAN X-Stream service, an upgrading of the speed of BGAN service, increasing its streaming rate of at least 50 per cent. Initially available in the Asia-Pacific region, BGAN X-Stream was developed as an enhancement to BGAN's other streaming services, as it guarantees streaming IP data rates from 384 kilobits per second (kbps) to nearly 500 kbps.[55]

Other examples of the important use of BGAN terminals are:

- support for a University of California archeological survey in remote Mongolia for the tomb of Genghis Khan;
- saving wildlife with conservationists from a rescue centre in the Indonesian jungle;
- research into climate change from a US-Russian project at the Northeast Science Station in Siberia;
- four-month journey across southern Africa to promote key conservation areas;

[54] "About Inmarsat – Our satellites," Inmarsat, accessed Apr. 12, 2013. Available: http://about.inmarsat.com/satellites.aspx?top_level_id=3&language=EN&textonly=False.

[55] "Intelsat Launches Faster BGAN Service," *Satellite Today*, accessed Apr. 24, 2009. Available: http://www.satellitetoday.com/st/headlines/30769.html.

- compilation of critical data on 2 million sq km of unchartered land within the Amazon rainforest;
- use by severe weather reporters in parts of the United States to relay dramatic early-warning live footage of tornados and hurricanes to local residents; and
- transmission of registration data for ID cards for Angolan citizens from remote rural areas to data centres for processing as part of a new national identity (ID) card project.[56]

To further its efforts in the mobile market and compete with existing satellite telephones from Iridium, Globalstar and Thuraya, Inmarsat began marketing, in 2010, the Isatphone Pro, which is the first hand-held satellite phone to be developed by Inmarsat and its use had been made possible by the successful upgrade of the ground network. It was designed mainly for professional users in the government, media, aid, oil and gas, mining and construction sectors. It offers voicemail, text, email messaging and location data for the user to look up or send by text.[57]

The phone had been slow to make it to market due to delays in Inmarsat's third and final Inmarsat-4 satellite and problems with the prime contractor for the phone, EMS Technologies of Norcross, Georgia. Sasken Communication Technologies Ltd. of Bangalore, India, had subsequently resumed the production of the phone. Inmarsat's handsets are linked to its three Inmarsat 4 satellites in geostationary orbit over the equator. Users are thus provided with an unbroken coverage link worldwide, except for the Polar Regions.[58] While recognising that the satellite telephone market was not a high growth business, with a rate of growth of only 3 to 5 per cent a year, Inmarsat could nonetheless prove to be a viable competitor in this area, especially since other operators, including Iridium and Globalstar, faced major cash demands to fund second-generation constellations in an unfavourable capital market.

Inmarsat and IMSO are involved in the new system, LRIT. The system was set up by IMO, through amendments to the SOLAS of 1974. It aimed to provide a global system for the identification and tracking of ships that extends the monitoring of ships beyond short range coastal networks. The

---

56 Inmarsat plc, *Annual Report* (2009): 29.

57 Ibid., 3.

58 Inmarsat, Services BGAN. Available: http://www.inmarsat.com/services/bgan.

system has become mandatory for all passenger ships, high speed craft, mobile offshore drilling units and cargo ships of over 300 gross tonnes. The system specifies that Flag States shall ensure that, as a minimum, four position messages per ship (every 6 hours) are sent, stored and available for designated entities entitled to have access to LRIT information. These entities include:

- Flag States, which may require information on the location of their vessels irrespective of the location;
- Coastal States, which may request information on ships up to 1,000 nautical miles from their coasts, irrespective of their flag;
- Port States, which may request information on those ships that have declared to have one of their ports as destination, irrespective of their location or flag; and
- Search and Rescue Authorities.

The system works by ships sending LRIT position information, derived from satellites, via Inmarsat or other telecommunications satellites, to a telecommunications provider and thence to an LRIT Data Centre, and to national control centres. One of the purposes of the system has been to enhance security for government authorities. For example, LRIT provides ship identity and current location information in sufficient time for the government to evaluate the security risk posed by a ship off its coast and to respond, if necessary, to reduce the risk.

An International LRIT Data Exchange (IDE) routes LRIT information between LRIT Data Centres. The European Union (EU) has also set up an EU LRIT DC which is the largest data centre of the international system, tracking by 2009 around 10,000 ships and generating at least 40,000 position reports per day. IMSO amended its Convention to enable the organisation to undertake the functions and duties of the International LRIT Coordinator for the establishment and operation of the international LRIT system. IMSO was subsequently appointed to that role by IMO. IMSO has hence signed LRIT Services Agreements with a number of States for the operation of National LRIT Data Centres.[59]

---

[59] See also IMO, Resolution MSC 275(85) of the *Maritime Safety Committee of the International Maritime Organisation (IMO)*, adopted by the MSC on Dec. 5, 2008, and information on the European Maritime Safety Agency and the EU Long Range Identification and Tracking of Ships Data Centre at www.emsa.europa.eu.

Inmarsat has furthermore benefitted from the fact that, unlike land-based customers, maritime and aeronautical subscribers tend to keep their equipment as long as they operate ships or aircrafts. As for aeronautical services, Inmarsat's business has grown significantly, a reflection of commercial airlines' introduction of in-flight communications services for passengers and crew. In the first quarter of 2010, aeronautical revenue was up nearly 36 per cent, and, similarly land-mobile revenue was up 31 per cent.[60]

Safety at sea has remained a primary commitment for Inmarsat Global, which is still the only approved provider of satellite communications for the GMDSS.[61] That service is used throughout the world by mariners and by the RCCs that are integral to ensuring that a safety alert is acted upon. The Inmarsat C SafetyNET service has been used for several years by the International Maritime Bureau to provide vital updates on reported pirate activity to approximately 70 per cent of the global fleet. The service enabled ship masters to access reports of pirate movements, giving them information to know which regions to avoid with high pirate activity and allowing them to re-route if necessary.

Disaster relief has long been an indispensable feature of Inmarsat's land mobile services. The Company continues to support TSF through direct funding and provision of free satellite terminals. TSF is able to reach disaster areas quickly and help charities and other aid organisations with their communications needs. During 2009, crises in many countries, including the Philippines, China, Indonesia, and in 2010, in Haiti and Chile, benefited from TSF's help and its use of Inmarsat communications facilities.

Although maritime revenue was flat for the first quarter of 2010, this should be viewed as a temporary decline given the number of new maritime satellite terminals activated during the period.[62] Shipping still accounted for more than half of Inmarsat's revenue and in general its maritime revenue has also remained strong, despite the general decline in new shipbuilding programs and the fact that some maritime fleet owners have reduced the number of ships in service, which deprives Inmarsat of service revenue. Increased market uptake of Inmarsat's newer FleetBroadBand maritime communications services have, however, offset this decline in ships as older vessels are being refitted with updated

60 Inmarsat plc Interim Management Statement, May 11, 2010. Available: www.inmarsat.com/cs/groups/inmarsat/documents/assets/018568.pdf.

61 Inmarsat plc. *Annual Report* (2009): 27.

62 Ibid.

Inmarsat equipment.[63] The FleetBroadBand 150 terminal offers vessels high-quality global telephone connectivity and simultaneous IP data at speeds of up to 150 kbps. For ships with FleetBroadBand equipment but no GMDSS facilities, the new 505 Emergency Calling service enables calls for help to be made immediately, anywhere in the world.[64] In furtherance of these services, in October of 2009, Inmarsat launched its Enhanced Super Quiet Time (ESQT) calling services to provide lower cost calling for crew abroad vessels that employ the FleetBroadBand service.

Inmarsat has also pursued a new mobile satellite services system in Europe, as it was selected by the EU in May 2009, along with Dublin-based Solaris Mobile, a joint venture between Eutelsat and SES Astra, to establish a next-generation mobile satellite communications system across the continent. Inmarsat's S-band satellite program, called EuropeSat, aims to deliver telecommunications services across all 28 EU member states by means of a hybrid satellite-terrestrial network for which Inmarsat is allying with partners. Currently, Eutelsat and SES are evaluating how and whether to proceed with Solaris Mobile since it has been struggling because of a defect with the S-band antenna on its satellite, while Inmarsat continues to seek a strategic partner willing to co-vest in the S-band venture.

Inmarsat has also linked up as a commercial partner with the European Space Agency (ESA) to become the commercial operator of a new satellite called Alphasat, to provide services over Europe, the Middle East and Africa. This venture will enable Inmarsat to develop new technology for its next generation network, and also provide network redundancy that will further enhance its services. The new satellite is expected to be launched on the Arianespace vehicle from French Guiana in 2013. In parallel with the Alphasat venture, Inmarsat has agreed to a co-funding opportunity with ESA for development of a range of new and enhanced Inmarsat products and services. Indeed, the process for building and launching a new satellite constellation is a multi-year venture, involving a business plan, availability of capital investment, choice of the procurement format choice of satellite, alternatives for increasing bandwidth and much else.[65]

---

63 Peter de Selding, "Inmarsat Sees Big Boost In Revenue." *Space News* 20, no. 19 (2009): 4.

64 Inmarsat plc. *Annual Report (2009)*: 27.

65 Inmarsat plc, *Annual Report* (*2009*): 8; and David Sagar, "Contractual and Procurement Issues in the Space Sector," *Proceedings of the 14th Summer Course of the European Centre for Space Law.* (Terni, Italy: European Centre for Space Law, Sept. 2005): 4–17.

Inmarsat is, however, anticipating a loss of some of its leased line traffic as the US Navy transitions from Inmarsat's L-band fleet services to wideband X-, C-, and Ku-band services. While Inmarsat expects its L-band services will still be employed by the US Navy, indeed other navies have increased their use of the services, it is focusing on adapting to the US Defense Information Systems Agency's (DISA) Future Comsatsom Services Acquisition program, which replaced its current resellers contract in 2012. Inmarsat seeks to compete and provide services across all three tiers of the new contract: pure bandwidth, subscription-based services, and the indefinite-delivery, indefinite-quantity aspect for managed network services.[66]

Inmarsat's sound financial standing will be further strengthened with its new distributor agreement, which finally altered an anachronistic distribution structure, a remnant of its pre-privatisation days, under which a limited number of "gatekeepers" had the ability to provide Inmarsat services directly to end users. The perpetuation of that structure was mandated by former Signatories as part of the Inmarsat privatisation process in order to preserve their historical exclusivity. Private equity firms, however, were key beneficiaries of this history, as Apax Partners acquired the France Telecom MSS business in 2006 and then acquired Telenor Satellite Services the following year, incorporating the two into its Vizada brand, as together the two distributors had been responsible for some 40 percent of all Inmarsat services worldwide. Thus, despite horizontal consolidation among Inmarsat distributors, which was arguably inconsistent with the goals of the ORBIT Act to promote a competitive global market for satellite communications services, these restrictions persisted in the form of contractual limitations until they expired in April 2009. More recently, Vizada again changed hands, having been acquired by Astrium in December 2011.[67]

### 11.2. *New Distributor Agreement and Recent Acquisitions*

In mid-April 2009 Inmarsat concluded many months of negotiations with its service distribution partners which resulted in the signing of a new five-year distribution agreement that shifted the balance of power back

66 Ibid.

67 "Astrium and Vizada become a world leader in satellite communications services," Astrium, accessed April 13, 2013, www.vizada.com/Satellite/Newsroom/News/Newsdetail/Newsdetail-page?page_id=7725.

toward the wholesaler Inmarsat. Due to an old formula that granted price reductions to a distributor once that said distributor achieved a set annual sales level coupled with mergers among its distributors, by 2008, Inmarsat was losing approximately US$ 65 million or nearly 10 per cent of its revenue in annual volume discounts, principally to Vizada and Stratos Global. The new agreement resulted in US$ 65 million in annual volume discounts being divided into thirds, with one third returning to Inmarsat's treasury over the next two years; another third employed to reduce pricing of services, as Inmarsat, which has maintained near uniform pricing worldwide, considers regional pricing schemes; and the final third paid to distribution partners under a formula that spreads the rewards throughout the chain of distributors, of which Inmarsat currently has some 28, though it is now free to license new distributors.[68]

Although Inmarsat technically had the right to appoint additional distributors for its services prior to the new distribution agreement, it was severely constrained from doing so by significant artificial barriers to entry, dictated primarily by the exclusivity arrangements imposed by the former Signatories. Any potential new distributor of traditional Inmarsat services had to invest in the construction and operation of an expensive gateway earth station facility that would "land" these services, as well as meet other threshold qualification criteria. A qualifying entity could not employ a common telecommunications practice and simply contract for access to an existing gateway. Thus, eight years after Inmarsat was converted from an intergovernmental organisation to a commercial enterprise, it had yet to appoint a single new distributor for its traditional services.[69] Indeed, even when Inmarsat sought to appoint distributors for other new services provided over the Inmarsat-4 network, it faced legal challenges from Telenor and France Telecom, as noted, now Vizada.

When the contractual restrictions in Inmarsat's distribution agreement expired in April 2009 and the new distribution arrangement was signed, Inmarsat was then allowed to assume full control of Stratos Global, a major distributor of Inmarsat services, which Inmarsat had purchased in December 2007 through a Canadian company established for that purpose, in part to reduce distributor resistance to the new agreement. Thus,

---

[68] Peter de Selding, "Inmarsat Comes Out Ahead in Revamped Distributor Deal," *Space News* 10, no. 16 (2009): 7.

[69] FCC Report to Congress As Required by the ORBIT Act, 26 and 27.

following approval by the FCC and the Canadian Minister of Industry, Inmarsat completed its acquisition of Stratos Global, which became a wholly owned operating division of Inmarsat managed by the existing Stratos management team. The vertical integration of Stratos arguably provides some significant public interest benefits, consistent with the competitive goals of the ORBIT Act, since vertical integration can reduce transaction costs, internalize incentives, take advantage of technological economies, and potentially reduce prices in the downstream market. As part of the terms of the arrangement, Inmarsat has segregated its wholesale operations from Stratos operating divisions and remains committed to a primarily indirect distribution model through its existing channels to market.[70] Inmarsat has stated it will treat its distributors equally, not favouring its new Stratos subsidiary nor Vizada, which accounts for nearly 40 per cent of Inmarsat's wholesale revenue.[71]

Inmarsat also chose Satcom Direct as a new distribution partner for its SwiftBroadBand services and the company also acquired a 19 per cent stake in the Canadian-based SkyWave Mobile Communications (Skywave), an investment which includes agreements for long-term global distribution and new product development as Inmarsat seeks to increase its presence in Canada.[72] Inmarsat also plans to enter the asset tracking services market through this acquisition of SkyWave.[73]

Inmarsat has entered into a partnership with SkyWave to expand its Satellite Low Data Rate (SLDR) or telemetry capabilities to complement its focus on the high-speed data market with products such as the broadband services. A telemetry device is typically an Inmarsat unit coupled with a GPS unit. The GPS unit records the position of the asset, such as trucks, containers, and high-value mobile assets like heavy plant machinery, and passes the information to the Inmarsat unit in the device for transmission to the customer via the Inmarsat network. The telemetry market offers the

---

70 "Inmarsat Completes Acquisition of Stratos, Signs New Distribution Agreements," Satellite Today, accessed April 12, 2009, www.satellitetoday.com/st/topnews/30638.html.

71 De Selding, "Inmarsat Comes Out Ahead in Revamped Distributor Deal."

72 "Inmarsat Acquires 19 Percent in Canadian Mobile Company," Satellite Today, accessed April 12, 2013, www.satellitetoday.com/st/topnews/30512.html.

73 Turner Brinton, "Profile Rebecca Cowen-Hirsch, President, Inmarsat Government Services," *Space News* (2010), accessed April 13, 2013. Available: http://www.spacenews.com/article/profile-rebecca-cowen-hirsch-president-inmarsat-government-services#.UWmIKCs9znF.

strategic benefit of bringing more traffic onto the existing network without material impact on costs or capacity for other services.[74]

To bolster its position in US government markets, in November 2009, Inmarsat announced its purchase of Segovia Inc., which provides integrated communications services, primarily to US government agencies.[75] Initially Inmarsat agreed to pay US$ 110 million with cash, but structured the deal to potentially pay additional amounts depending on Segovia's performance over three years, that is, through 2012.[76] Segovia designs, deploys and operates mixed satellite and terrestrial networks and is a leading provider of secure internet protocol (IP) managed solutions and services to various US government services. The acquisition will thus enable Inmarsat to improve its relationship with key US government customers, notably the Department of Defence.[77]

Through these and other acquisitions, the new distributor agreement, and its continued innovative efforts to enhance the flexibility and mobility of its services, Inmarsat has strengthened its ability to compete with other mobile satellite-constellation operators such as Orbcomm, Iridium and Globalstar as well as Fixed Satellite Service (FSS) providers. The ORBIT Act specified a number of criteria for determining whether Inmarsat's privatisation was "pro-competitive," as its stated purpose sought to promote a fully competitive global market for satellite communications services. With the new distributor agreement and the consequent removal of the contractual limitations in the former distribution agreements, Inmarsat may be better able to facilitate the actualization of this US legislative goal, but the industry has in fact witnessed increased consolidation, with, of course, the approval of the FCC.

### 11.3. *Global Xpress*

Global Xpress will represent Inmarsat's first foray into frequency bands outside of the conventional MSS bands. involving the launch of three high-powered Ka-band satellites starting in late 2013, with full global connectivity

---

74 Inmarsat plc. *Annual Report* (2009): 7–8.

75 "Inmarsat completes acquisition focused on government services," Inmarsat, accessed April 12, 2013, www.inmarsat.com/corporate/media-centre/press-releases/inmarsat-completes-acquisition-focused-on-government-services.

76 Ibid.

77 Inmarsat plc. *Annual Report* (2009): 8.

expected by the end of 2014. The service will be delivered on three newly-procured Inmarsat-5 satellites, which will be built by Boeing, based on its flight-proven, state-of-the-art 702HP satellite bus, and launched on Proton launch vehicles. The estimated cost of the Inmarsat-5 and Global Xpress programme will be US$1.2 billion over 4.5 years, inclusive of space segment (satellites, launches and insurance), additional ground network infrastructure as well as product development.[78]

According to Inmarsat, Global Xpress is expected to revolutionise broadband services in the maritime world, while maintaining Inmarsat's current commitment to the outstanding quality, global coverage and seamless mobility that its customers have come to expect. The Inmarsat-5 satellites will provide customers with a comprehensive range of global mobile satellite services, including mobile broadband communications for deep-sea vessels, in-flight connectivity for airline passengers and streaming high-resolution video, voice and data. As such, Global Xpress is intended to represent the next stage in the evolution of remote and mobile communications, supporting broadband connections at speeds comparable with terrestrial networks, delivering seamless global coverage and mobile broadband with downlink speeds of up to 50Mbps to customer terminals from 20cm-60cm in size. Inmarsat does not, however, envision Global Xpress as a replacement for its current L-band service offerings, but rather will allow customers to take advantage of unique hybrid packages that use both its L-band and Global Xpress networks.[79]

## 12. *Conclusions*

Privatising Inmarsat was a lengthy process—indeed it took longer for Inmarsat to become the first intergovernmental organisation to be transformed into a private company than for the States to draw up the original Convention and Operating Agreement in the 1970's. Hindsight suggests that the outcome was inevitable, and that the forces seen dimly in 1990 gathered pace and left no realistic option by 1998. This assessment is

[78] "Inmarsat Global Express," Inmarsat. Available: www.igx.com. *See also* Inmarsat, Inmarsat announces $1.2b investment in next generation Ka-band satellite network, Aug. 6, 2010. Available: www.inmarsat.com/corporate/media-centre/press-releases/inmarsat-announces-1.2bn-investment-in-next-generation-ka-band-satellite-network.

[79] "Inmarsat Global Express," Inmarsat. Available: www.igx.com.

borne out by the fact that INTELSAT and EUTELSAT undertook comparable restructuring programmes in the years that followed, drawing to some extent on INMARSAT's experience.

From its start of services in 1982 up till the post-2000 world, INMARSAT and its successor has had remarkable success in developing the maritime satellite telecommunications market, and extending its services extensively in the land-mobile market, especially in remote areas. Substantial progress has also been made with its aeronautical services where its full potential, especially with regard to passengers, is yet to be achieved.[80] A decentralized distribution network has enabled it to serve many thousands of mobile users and in turn generate substantial traffic and revenues, in which the volume of data traffic has grown larger than voice. Several successive generations of satellites, each incorporating new technology and increased communications capacity, serve an increasing variety of user applications through a wider range of terminals.

INMARSAT, and its successor, has kept pace with, and responded well to worldwide political, economic and technological evolution throughout the decades, in particular the liberalisation and privatisation of the global telecommunication industry. Its continuing commercial success as a privatised entity will depend on how it faces up to increasing competition, and its ability to capture the mass markets of millions of users. Its new BGAN service and prospects for expanding aeronautical passenger usage will be a vital element in these aims. Successful negotiation of radio frequency (RF) spectrum, always in short supply and high demand, will be vital, as will favourable intersystem co-ordination with other operators. Other factors will be the provision of seamless interoperability, a single telecommunications service provider, a single customer service centre and one billing system.

Maintenance of public service obligations, so crucial to the original intergovernmental enterprise, is severely challenged in the new privatised, profit-motivated company. Inmarsat has to date ensured the provision of public services such as safety of life, disaster relief, security, and peacekeeping. Indeed, use of its system to keep the public informed in conditions when all other forms of communication have broken down has revolutionised international response in events of disaster. Long distance road and

---

80 Ahmad F. Ghais, "*Inmarsat: A Success Story*," in *Success Stories in Satellite Systems*, ed. D.K. Sachdev (Virginia: AIAA, 2009): 95–119.

rail transportation have also benefitted enormously by these facilities. Communications for the worldwide shipping industry have been transformed dramatically by the replacement of wireless by satellite communications for commercial and personal purposes and, most importantly, for maritime safety through the GMDSS. The company has, however, succeeded in removing some of these obligations from its Public Service Agreement with IMSO. Whilst the continued provision of maritime safety services is assured under the Public Service Agreement, the maintenance of the other public services will no doubt depend mainly upon their profitability.

The future commercial success of the Company, and thus its ability to maintain the maritime distress and safety services and other public services, will not depend just on continuing to provide the existing range of maritime, aeronautical and land mobile services, but a business strategy to pursue a range of new data opportunities. Although Inmarsat has no immediate need for new spacecraft, the company is exploring a range of options for its next-generation global satellite system, including advanced L-band, Ka-band and other frequencies as it develops strategic plans for its future in mobile broadband and possible alternatives for a fifth-generation system. Long term service expansion in conjunction with continued flexibility and mobility of its services is critical to Inmarsat's future viability, especially as the company faces substantial competition from several global and regional mobile satellite service providers as well as increased competition from the fixed satellite services industry and a broad array of terrestrially based technologies. Inmarsat's restructuring experience will help to forge a new chapter in international cooperation, based on a constructive relationship between the private sector and governments in the provision of space telecommunications to the world community.

# Chapter Three

## Intelsat: Pre and Post-Private Equity Ownership

*Patricia K. McCormick*

### 1. *Introduction*

In November 2000, the 143 Member States of INTELSAT[1] made the historic decision to restructure the organisation into two separate entities. The actual satellite system of INTELSAT was privatised by transferring substantially all of its assets and liabilities, including satellites and orbital fillings, to a new company established for this purpose, Intelsat, Ltd., incorporated in Bermuda, and its subsidiaries.[2] The other entity, which retained the name International Telecommunications Satellite Organisation (ITSO), is a highly modified version of the intergovernmental organisation. With an initial lifetime of twelve years, which was extended through at least 2021, ITSO is designed to supervise and monitor the private company to ensure that it meets its public service and lifetime connectivity obligations to those developing countries largely dependent on the services and pricing mechanisms of Intelsat. The privatisation of this instrumental international satellite organisation has fundamentally changed its mission to one of an international regulatory body overseeing international public service telecommunications services.

---

1 INTELSAT was originally established by means of the Agreement Relating to the International Telecommunications Satellite Organisation (INTELSAT)(hereinafter INTELSAT Agreement), done 20 Aug. 1971, entered into force 12 Feb. 1973, as amended 17 Nov. 2000, amended version applied provisionally 18 July 2001, entered into force 30 Nov. 2004, 1220 U.N.T.S. 21, 23 U.S.T. 3813, T.I.A.S. 7532.

2 Please note that henceforth in this document, INTELSAT shall refer to the satellite organisation prior to privatisation and Intelsat, Ltd. or simply Intelsat will refer to the private operator and its subsidiaries after its privatisation.

This chapter[3] examines the ramifications of the structural alterations of this organisation, which especially affect developing countries, since more than 90 developing countries remain largely dependent on Intelsat for their telecommunications needs. It begins with a review of the historical development of INTELSAT, including a critical analysis of relevant US legislation, such as the Communications Satellite Act of 1962 (Satellite Act)[4] and the subsequently-enacted Open-market Reorganisation for the Betterment of International Telecommunications Act (ORBIT Act),[5] which in accord with the expanded embrace of neo-liberalism and agreements of the World Trade Organisation (WTO), have affected significant changes within the satellite industry, including the privatisation of INTELSAT.

Additionally, this chapter assesses the effectiveness of ITSO in addressing issues related to universal access, since the existing inequities in the global distribution of power and wealth can grow increasingly entrenched in the absence of effectual international access safeguards. The mechanisms previously employed by INTELSAT essentially used the profits from the lucrative, high-density international traffic routes to subsidize service in rural and economically deprived or low-density route areas. While the efficacy of this subsidization process has been questioned, most developing countries, with such notable exceptions as Brazil, China, India, and Indonesia, do not have the indigenous capabilities for developing satellite technology and have difficulties in paying transponder fees and acquiring other commercial satellite services.[6] The obstacles facing developing countries remain unresolved as commercial imperatives prevail over development concerns in a market driven environment. Finally, this chapter considers a number of developments following INTELSAT's privatisation and their impact on the continued availability of universal access to satellite services.

---

3 Please note that this chapter is based, in part, on the previously published work, "The Privatisation of INTELSAT: The Transition from an Intergovernmental Organisation to Private Equity Ownership," in *Telecommunications Research Trends*. Ed. H. Ulrich and E. Lehrmann (New York: NOVA Science Publishers, Inc., 2008).

4 Communications Satellite Act (hereinafter Satellite Act), 47 U.S.C. § 701 (1962).

5 Open-Market Reorganisation for the Betterment of International Telecommunications Act (hereinafter ORBIT Act), 47 U.S.C. § 761 (2000).

6 Daya Kishan Thussu, "Lost in Space," *Foreign Policy* 124 (May/June 2001): 71.

## 2. *Intelsat – The Early Years*

### 2.1. *The US Communications Satellite Act of 1962*

The United Nations General Assembly, in recognition of the beneficial applications of satellites and in response to the 1960 US launch of the first telecommunications satellite, Echo I, for switched voice transmissions, which was followed by Telstar and Relay in 1962 for broadcast quality transmissions, passed a resolution in 1962 that called for a system of 'communications by satellite' which would be "available to the nations of the world ... on a global and nondiscriminatory basis."[7] President Kennedy thus envisaged a global satellite network that would aid developing countries and rely on considerable involvement of government agencies, instead of being simply an organ of private enterprise. Although the final version of the Satellite Act was largely a comprised version of the legislation originally envisioned by Kennedy, it did represent a major pillar in US Cold War foreign policy objectives, as it positioned the United States, rather than the Soviet Union, as the principal benefactor in bringing the benefits of this nascent technology to the rest of the world. The inception of the satellite industry in the United States, however, pitted public advocates and private businesses against one another in a struggle over its control, as occurred throughout a series of Congressional hearings in 1961–1962. The resultant Satellite Act represented hard-fought concessions on behalf of all legislators as it was filibustered for weeks before cloture was invoked.[8] The Satellite Act, however, essentially represented a victory for private interests as it culminated in significant participation of the private sector in the emerging satellite industry, since it authorised the formation of the Communications Satellite Corporation or COMSAT to serve as the United States' chosen instrument for overseeing the development and global promotion of this emerging industry.

Senators Wayne Morse (D-Oregon) and Carey Kefauver (D-Tennessee) and Congressman Emanuel Celler (D-New York) formed an alliance against the private ownership of COMSAT in conjunction with the American

[7] Michael A. Einhorn, "Restructuring and competition in international telecommunications: the case of INTELSAT," *Information Economics and Policy* 10, no. 2 (June 1998): 198.

[8] Joseph N. Pelton and Burton I. Edelson, "Reinventing INTELSAT and INMARSAT," *Aerospace America* 35, no. 1 (January 1997): 28.

Communication Association, a union of telecommunications workers, as well as Assistant Attorney General Lee Loevinger and renowned academic scholars, Herbert Schiller and Dallas Smythe.[9] This alliance voiced concerns that a privately owned COMSAT would strengthen the private sector's control over public airwaves. They called for increased public participation in the congressional hearings as well as government ownership of the satellite company. Senator Robert Kerr (D-Oklahoma), in staging an opposing alliance with major communications corporations, including RCA and AT&T, the latter who ostensibly sought to extend their telephony monopoly to satellite communications, proposed a bill that called for the privatisation of satellite communications, since space communications, as Kerr contended, offered beneficial new business opportunities for the private sector.[10]

Amidst this controversy of competing pressures, COMSAT emerged as a compromise. COMSAT was incorporated as a publicly traded company in 1963, though the Act permitted up to 50 percent aggregate ownership by then-existing common carriers. Federal agencies, including the State Department, the Federal Communications Commission (FCC), the National Aeronautics and Space Administration (NASA), and the Office of Telecommunications Policy (now the National Telecommunications and Information Administration (NTIA)), and the President were each charged with monitoring duties and the President was granted the authority to appoint three directors to the COMSAT Board. The organisation, though designed to operate as a private business while being regulated to serve the public interest, historically favoured the corporate interests of its mandate as it straddled the often conflicting objectives of private enterprise and the public good.

Since building a global telecommunications system was thought to be beyond the capacity of any single country or private entity, COMSAT was charged with raising private financing and securing foreign partners to develop a global satellite communications system. Although US negotiators initially considered creating a new global satellite system through a series of bilateral agreements, the Committee on European Post and Telecommunications would only accept a multilateral agreement.[11] Two

---

[9] "Communications Satellite Corporation," Museum of Broadcast Communications. Available: http://www.museum.tv/archives/etv/C/htmlC/communication/communication.htm.

[10] Ibid.

[11] Pelton and Edelson, "Reinventing INTELSAT and INMARSAT," 29.

years of international negotiations thus ensued which resulted in August 1964 in the formation of the International Telecommunications Satellite Cooperative (INTELSAT) as the world's first commercial[12] satellite operator, originally comprised of just eleven member nations, which thereafter established the first global satellite communications system. At the time INTELSAT was considered a major innovation in the domain of international organisations since its structural organisation, international procurement procedures, contractual services, and operations were modelled on a commercial corporate structure, not the United Nations system.[13] Ownership and voting rights, specifically in the Board of Governors, were related to usage of the system, with the United States guaranteed a dominant role since COMSAT was the largest investor.

COMSAT was also initially responsible for the technical management of the system. In 1965, COMSAT and what had then grown to an ad hoc partnership of 44 nations successfully launched into geostationary orbit the first commercial communications satellite, "Early Bird." Several more geosynchronous satellites were launched in subsequent years. In 1971, agreement was reached on converting the cooperative into a more formal treaty-based international organisation comprised of 86 nations, to be known as the International Telecommunications Satellite Organisation, with the transformation taking effect in 1973.[14] The special legal status of INTELSAT granted it immunity from paying taxes to any national government as well as immunity from jurisdiction, which prevents courts from considering lawsuits against INTELSAT; archival and testimonial immunity, which protects INTELSAT from being forced to submit documents or testimony of its employees; and immunity of assets, which precludes courts

12 INTELSAT was from its inception a hybrid organisation in that there was both a commercial and an intergovernmental aspect to it. While on the commercial side, INTELSAT functioned as a cost-sharing cooperative with investment tied to usage, the individual Signatories clearly were seeking to generate profits as a result of their use of the INTELSAT system. Moreover, to the extent that there was a complex mechanism in place that would allow certain Signatories to have investment levels lower than their utilization, and other Signatories to have investment levels higher than their utilization, there was a profit element included in the pricing structures for each of the services INTELSAT offered. The profit motive was in fact in place prior to privatisation.

13 Pelton and Edelson, "Reinventing INTELSAT and INMARSAT," 29.

14 "Our History 1970s – A Decade of Expansion," Intelsat, Ltd., (2003). Available: http://www.intelsat.com/aboutus/ourhistory/yr1970s.aspx.

from enforcing monetary judgments against INTELSAT.[15] Its treaty status also assisted in ensuring its access to the domestic markets of member states, a valuable privilege or asset not readily conferred upon the entry of its competitors in forthcoming years.

## 2.2. *Governance*

Operating as a user-owned cooperative, INTELSAT had a multi-tiered governance structure. This configuration, reflective of INTELSAT's dual roles as an intergovernmental treaty organisation and a commercial provider of telecommunications services, consisted of the Assembly of Parties, the Meeting of Signatories, the Board of Governors, and the unitary Executive Director General. The Assembly of Parties represented the principal political organ of INTELSAT, the pinnacle of its structure, as it comprised each country that was party to the INTELSAT Articles of Agreement, which, at the time of its privatisation, constituted 143 member states with eight Western countries accounting for half the controlling shares and the United States holding the largest investment of 20.4 per cent.[16] While each country had one vote at the Assembly of Parties, which met biennially to establish general policy and long-term objectives and was needed to ratify major decisions, this body generally followed the counsel of the Board of Governors, which consisted of Signatories to the Operating Agreement.[17]

Each member state designated a single telecommunications entity to be a Signatory to INTELSAT. Each Signatory was responsible for the requisite capital contributions to finance its state's share of INTELSAT, and to conduct the obligatory commercial and technical operations to provide the transmission capacity and communications services of the satellite system to carriers and users in their respective nation.[18] Generally the government-owned and operated Post, Telegraph and Telephone (PTT)

---

15 Kenneth Katkin, "Universal Global Interconnection After INTELSAT," (presented at the 30th Research Conference on Communication, Information and Internet Policy, Washington, D.C., Sept. 29, 2002): 11.

16 Thussu, "Lost in Space," 71.

17 Robert W. Hahn and Randall S. Kroszner, "Lost in Space US International Satellite Communications Policy," *Regulation The CATO Review of Business and Government* 13, no. 2 (Summer 1990): 58.

18 Katkin, "Universal Global Interconnection After INTELSAT," 5.

monopolies were so designated, though in the case of the United States, COMSAT served as its Signatory to INTELSAT. COMSAT, like the other Signatories, then had the sole right of access to INTELSAT. The long distance telephone service providers, news organisations, private network operators, and radio and television broadcasting networks could not lease services directly from INTELSAT, but were required to go through its state's Signatory if it sought to employ international satellite services.[19] In reselling capacity to customers, these Signatory monopolies could thus overprice international service and restrict competitive entry.[20]

The structural organisation of INTELSAT further secured the interests of these monopolies, since designated representatives from each Signatory body comprised the Meeting of Signatories, which, like the Assembly of Parties, adhered to the one-country-one-vote rule. The Meeting of Signatories met annually to, among other duties, establish rules governing the rates and adjustments of INTELSAT satellite transmission capacity, adjust capital contribution ceilings, authorise new earth stations, consider the financial implications of proposed programs, and resolve disputes among member states.[21]

The general, day-to-day operation and administration of INTELSAT, however, resided with the Board of Governors or those who most used the system, since representation on the INTELSAT Board of Governors was apportioned in accord with the Signatory's share ownership, itself based on each Signatory's utilization of INTELSAT satellite transmission capacity.[22] Since the principal responsibility "for the design, development, construction, establishment, operation and maintenance of the INTELSAT space segment" was borne by the Board of Governors, the stipulations regarding representation on the Board consequentially meant that those industrialized, hegemonic countries, specifically the United States and the United Kingdom, extended the greatest influence over the direction of the organisation.[23] Akin to the conflicting goals and agendas of developed and underdeveloped states in most international forums, conflicts arose in INTELSAT between the few wealthy investors that financed most of INTELSAT's

---

19 Hahn and Kroszner, "Lost in Space US International Satellite Communications Policy," 58.

20 Einhorn, "Restructuring and competition in international telecommunications," 200.

21 Katkin, "Universal Global Interconnection After INTELSAT," 6.

22 Hahn and Kroszner, "Lost in Space US International Satellite Communications Policy," 58.

23 Katkin, "Universal Global Interconnection After INTELSAT," 6.

investments and the majority of smaller Signatories that owned very little capacity. The hegemonic position of the major INTELSAT Signatories, such as those from the United States and the United Kingdom, is in accord with the work of Cowhey who contends that the domestic policies of the dominant, prevailing players are the primary source of international regime change.[24]

Directly accountable to the Board of Governors for the performance of all management functions was the fourth level of governance, the Director General, who exercised executive responsibility over daily operations.[25] This post was elected by the Board of Governors and confirmed by the Assembly of Parties. Several candidates were nominated from various geographic regions, and then there were successive rounds of balloting, in which the candidate with the lowest vote total each round was eliminated, until one candidate received a two-thirds majority of the vote or until there were only two candidates remaining, in which case the candidate with the higher vote total would then become the Director General-elect, to serve as the chief executive and legal representative of INTELSAT.

The quadripartite structure of INTELSAT, arguably, hindered its development and ability to compete in latter years. The administrative process for resolving the conflicting needs and agendas of its numerous Signatories regarding competitive positioning, tariff-setting, capacity additions, and technological updates, among other issues, was slow, cumbersome, highly political and costly to administer.[26] With the advent and subsequent growth of competing separate satellite systems, INTELSAT was thwarted in its ability to rapidly respond to competitive offerings that bundled products and services and featured price discounts. Since INTELSAT only provided space-segment services and neither owned nor operated earth stations and interconnecting terrestrial facilities, it could not market itself as an all encompassing service provider.[27]

---

24 Peter F. Cowhey, "The International Telecommunications Regime: The Political Roots of Regimes for High Technology," *International Organisation* 44, no. 2 (March 1990): 173–174.

25 Katkin, "Universal Global Interconnection After INTELSAT," 7.

26 Einhorn, "Restructuring and competition in international telecommunications," 211.

27 Johannes P. Pfeifenberger and Hendrik S. Houthakker, "Competition to International Satellite Communications Services," *Information Economics and Policy* 10, no. 4 (1998): 405.

## 2.3. *The Emergence of Separate Satellite Systems*

At the time of INTELSAT's creation, the prevailing economic rationale held that natural monopolies in the provision of public utilities were the most efficient. In virtually every country in the world, providers of telephony and telegraph services were either private monopolies, crown corporations, or, most commonly, government Ministries of Post, Telegraph and Telephone. This historical structure, in accord with the natural monopoly argument that contended it was more economically efficient to construct a single satellite network connecting different countries rather than building several competing satellite systems, gave rise to the formation of INTELSAT as a single monopoly provider of satellite services. The large initial costs or high sunk costs in building and operating satellites coupled with the relatively low cost of adding additional users in conjunction with the arguments about economies of scale and unnecessary duplication of costly facilities, contributed to the general assumption that competition in satellite communications was neither feasible nor desirable.[28] Maintaining INTELSAT's role as the premier, if not sole global satellite operator, also dovetailed with US foreign policy objectives and, thus, the United States itself was an early proponent of safeguarding INTELSAT's special and unique role from possible competing systems.

At the same time, though, potential competition from other systems was never intended to be totally foreclosed. Section 102 of the Satellite Act[29] expressly provided for the eventual creation of additional communications satellite systems as it did not "preclude the creation of additional communications satellite systems, if required to meet unique governmental needs or if otherwise required in the national interest." Although the accession of the United States to the INTELSAT Agreement may have in effect delayed the deployment of separate satellite systems serving the United States, on 28 November 1984, President Reagan, in accord with his administration's policies promoting deregulation, increased market competition, and less government involvement in commercial enterprises, which by then had overtaken foreign policy considerations, made a determination that separate international satellite systems were required in the country's national interest.[30] As Orion and the Pan-American Satellite Corporation

28 Hahn and Kroszner, "Lost in Space US International Satellite Communications Policy," 58.

29 Satellite Act; as amended 1978; *Space Law—Basic Legal Documents*, E.III.2.

30 Katkin, "Universal Global Interconnection After INTELSAT," 9.

(PanAmSat), in addition to four other companies, made their initial applications to the FCC, INTELSAT responded by encouraging a boycott, as it adopted a resolution on 31 January 1985, urging its members to refrain from participating in the construction of any separate international satellite systems not owned or operated by the PTTs.[31] Despite such resistance, the FCC proceeded to make orbital slots available and, with the launch of PAS-1 in 1988, PanAmSat became the first private satellite operator in direct competition with INTELSAT and it achieved full global coverage in 1994.[32]

Contrary to its strong opposition to private satellite systems, INTELSAT did not present such a defiant position against the European PTTs, which in conjunction with others, had initiated planning in the late 1970s for the creation of a European satellite system largely modelled on INTELSAT. The European Telecommunication Satellite Organisation, EUTELSAT, headquartered in Paris, was created in 1977 as an intergovernmental organisation to develop and operate a satellite-based telecommunications infrastructure for Europe.[33] Similarly, after studying the operational requirements for a satellite system devoted to maritime purposes, the Convention for the International Maritime Satellite Organisation (INMARSAT)[34] was adopted in September 1976. Hence, on 16 July 1979, INMARSAT, headquartered in London, effectively came into being as an intergovernmental organisation to improve maritime communications, and, thus, alleviate distress and emergencies and generally enhance safety of life at sea, increase the efficiency and management of ships, and provide maritime public correspondence services and other communications for the maritime community.[35] Initially, these systems were operated by INTELSAT Signatories, and

---

31 Hahn and Kroszner, "Lost in Space US International Satellite Communications Policy," 62.

32 Pfeifenberger and Houthakker, "Competition to International Satellite Communications Services," 408.

33 Art. III(a), Convention Establishing the European Telecommunications Satellite Organisation (EUTELSAT), done 15 July 1982, entered into force 1 Sept. 1985, as amended 20 May 1999, applied provisionally 2 July 2001, entered into force 28 Nov. 2002 (hereinafter EUTELSAT Amended Convention), K Misc. No. 4, Cmnd. 9154, *Space Law – Basic Legal Documents*, C.II.1., C.II.2.

34 Art. 27, Convention on the International Maritime Satellite Organisation (INMARSAT) (hereinafter INMARSAT Convention), done 3 Sept. 1976, entered into force 16 July 1979, as amended 1998, amended version applied provisionally 15 April 1999, entered into force 13 July 2001, 1143 U.N.T.S. 105, 31 U.S.T. 1.

35 Convention on the International Maritime Organisation (hereinafter IMO Convention), entered into force 17 Mar. 1958, 289 U.N.T.S. 489 U.S.T. 621. Available: http://www.imo.org/Conventions/contents.asp?doc_id=674&topic_id=257.

thus presented little direct competition. INTELSAT, however, sought to contain the growth and development of competing private satellite systems. Although the INTELSAT Articles of Agreement recognised that separate satellite systems might be of value in the future and Article XIV(d) specified procedures for the coordination and approval of separate satellite systems, this aspect of its charter proved to be an effective anticompetitive instrument as it enabled INTELSAT to assume a quasi-regulatory role in regard to its competition.

Article XIV(d) required potential competitors to coordinate and acquire approval for their activities from INTELSAT. The lengthy review process, which included assessing their competitors' business plans before beginning operation, enabled INTELSAT to potentially impede its competition or, at the least, delay its entry, a claim made, albeit unsuccessfully, in PanAmSat's antitrust suit against COMSAT, and, by extension, INTELSAT.[36] INTELSAT's ability to thwart its competition resided in its Charter's stipulation that separate satellite systems, in addition to meeting technical compatibility and non-interference criteria, essentially demonstrate that their system would not cause INTELSAT 'significant economic harm,' a relative and ambiguous concept since the calculation of such harm was not denoted. INTELSAT employed the ambiguity of 'economic harm' to its advantage as it discriminated in its calculation, impeding the entry of private satellite systems by subjecting them to an arduous approval process with harm calculated as the maximum traffic diversion, but more readily granting approval, due to a calculation of negligible economic harm, to those regional systems owned and operated by the PTT Signatories, including EUTELSAT, the Arab Satellite Organisation (ARABSAT) and Indonesia's Palapa system.[37]

The quandary, in essence, lay in delinking separate satellite systems from the existing public-switched networks, both international and domestic, from which INTELSAT derived the majority of its revenue. The Reagan administration devised a policy, the separate systems restrictions, that licensed new competitive systems, but limited their traffic to new video and business systems.[38] The separate systems restrictions, which initially prohibited private satellite systems from providing services that accessed the public switched network, were designed to protect INTELSAT from competition in its "core" market of long-distance telephone traffic and thus

36 Hahn and Kroszner, "Lost in Space US International Satellite Communications Policy," 61.

37 Ibid.

38 Pelton and Edelson, "Reinventing INTELSAT and INMARSAT," 30.

ensure subsidized service to developing countries. This provision was not only never applied to competing terrestrial networks, such as transoceanic fibre optic cables, but one can challenge the validity of the cross-subsidy in satellite circuits, arguing that INTELSAT's pricing policy was employed to impede competition rather that to subsidize the less developed nations.[39]

## 2.4. *Pricing: Concerns and Criticism*

INTELSAT's mandate to charge uniform prices for similar services was considered a subsidy to the thin-route users in developing countries, because prices were not related to the actual costs of service provision. One can, however, contend that INTELSAT's average costs were predatory because the system was overbuilt.[40] Since the cost of overbuilding the system could be allocated among all the users, high-density countries had no incentive to accurately estimate capacity needs and the manufacturers and launchers of satellites, principally located in industrialized states, pressured their governments to support high estimates to increase the demand for their products, with the end result being that users on low-density routes paid higher average prices.[41] The uniform pricing policy thus encouraged excess capacity, as the practice of geographically averaging tariffs, presumably to shield countries with low volume, generated uneconomic bypass and strand investment.[42] Excess capacity also enabled INTELSAT, as a protected monopolist, to undercut its competitors' prices and introduce new classes of services at a minimal additional cost.

While INTELSAT's prices were geographically non-discriminatory, tapered tariffs adopted over time provided discounts for both duration and volume.[43] Furthermore, INTELSAT's technological choices were more appropriate and cost-effective for high-density use and developing countries thus bore a very high ground segment cost per circuit for their low-density routes.[44] Despite the considerable capital investment of constructing large earth stations borne by developing countries, INTELSAT assumed no liability for any damages to the ground arrangements that

39 Hahn and Kroszner, "Lost in Space US International Satellite Communications Policy," 62.
40 Ibid, 62–63.
41 Ibid, 63.
42 Einhorn, "Restructuring and competition in international telecommunications," 200.
43 Ibid.
44 Hahn and Kroszner, "Lost in Space US International Satellite Communications Policy," 63.

contracting countries could suffer.[45] Arguably, in the absence of competition, a non-profit consortium arrangement should design tariffs to protect against exorbitant prices, but like other monopolies, INTELSAT enjoyed a substantial, if not excessive, return on its investments, averaging 14 to 18 per cent per year, protection of which may not be warranted.[46] This high rate of return, which only directly benefited members with net ownership over usage, served to establish the 'wholesale' rate for international communications levied by INTELSAT at a much higher rate that, in turn, created a basis for inflating national retail rates through a said Signatory.[47]

While the rate of return was arguably high, these arguments did fail to take account of the fundamental point that INTELSAT was, at its core, established to function as a cost-sharing cooperative, and the built-in rate of return did serve one laudatory purpose of providing a means whereby smaller countries could maintain their status as member countries without being forced to assume increased investment obligations as their relative utilization levels increased over time. In any event, the success of PanAmSat in negotiating with INTELSAT Signatory countries in South America and the Caribbean can be viewed as providing additional evidence against the efficacy of INTELSAT's service subsidization, though INTELSAT did provide personnel training and technical assistance to its Signatories, especially in developing countries, and thereby acquired an understanding of their needs through these interactions.

Although INTELSAT thus served as an institutional framework that facilitated information and technology transfer, its marketing and pricing arrangements allegedly grew outdated and the organisation was unable to contend with its new competitors.[48] As companies from countries throughout the world were able to build, launch, and operate satellites, what became of critical importance was a firm's sales and marketing ability, not one of INTELSAT's strengths. Furthermore, in concurrence with the emergence of competition from separate satellite systems, INTELSAT began to face substantial competition in the market for international communication transmission capacity from the proliferation of transoceanic fibre-optic

---

45 Juliana Gruenwald, "Panel Moves to Privatise Satellite Organisations," *Congressional Quarterly Weekly* 56, no. 12 (1996): 744.

46 Michael Einhorn, "Intelsat: A Reform Proposal," *Space Communications* 14, no. 2 (1996): 139.

47 Pelton and Edelson, "Reinventing INTELSAT and INMARSAT," 32.

48 Einhorn, "Restructuring and competition in international telecommunications," 210–211.

cables. Capable of delivering many of the same services as satellites, but at lower costs, the first transoceanic cable, extending across the Atlantic Ocean from the United States to the United Kingdom, was completed in 1988 and the first trans-Pacific cable became operational in 1991.[49] These technological developments, in conjunction with INTELSAT's wieldy politicized bureaucratic governance structure and questionable uniform pricing practice as an effectual subsidization mechanism, coupled with broader changes in the global environment in accord with the dictates of the WTO, since market changes do not occur in a vacuum but generally reflect policy changes, together provided impetus for the reform of the organisation.

## 3. *Intelsat – The Era of Restructuring*

### 3.1. *Resolution to Reform*

The goals of the reform were two-fold – to enable INTELSAT to more effectively market its services while simultaneously forfeiting uneconomic advantages and immunities its had henceforth enjoyed, since the privileges INTELSAT derived from its special legal status proved to be a source of strife among its private competitors.[50] Several options that would enable INTELSAT to successfully adapt to the new competitive environment were proposed and some were integrated into propositions to reform the organisation. These various alternatives included separating basic services from value-added and high-growth services; restructuring the tariff system to enable greater pricing flexibility, which could entail setting limits on charges or a price cap for thin routes and developing countries while other service rates could be commercially determined by the market; streamlining operations for maximum efficiency; acquiring other international sources of support for development activities; exploring strategic partnerships to develop new services and products; and restructuring the organisation's capital financing to allow new sources of investment.[51]

In regard to allowing non-Signatory users and investors, in 1992, INTELSAT introduced four new types or levels of direct access to INTELSAT satellites by non-Signatory carriers and users. Although the first two levels

---

49 Katkin, "Universal Global Interconnection After INTELSAT," 10.

50 Einhorn, "Intelsat: A Reform Proposal," 143.

51 Pelton and Edelson, "Reinventing INTELSAT and INMARSAT," 31.

were confined to the exchange of technical and operational information, levels 3 and 4 pertained to access to communication services. Level 3 direct access allowed customers to enter into a contractual agreement with INTELSAT for ordering, receiving and paying for INTELSAT space segment capacity at the same rates charged to INTELSAT signatories. Level 4 direct access enabled customers, in INTELSAT member states only, to make a capital investment in INTELSAT in proportion to its utilization of the INTELSAT system, as well as obtain INTELSAT space segment capacity at INTELSAT tariff rates, though rights to participate in INTELSAT's governance process were not generally accorded to a level 4 customer unless special arrangements were made by the Party and the official Signatory representing the country. Indeed, INTELSAT only allowed direct access in countries where it was authorised by the Signatory representative. As of 1999, level 3 direct access was available in 65 countries and level 4 in 29 countries for a total of 94 Signatory states, though, at that time, neither level was available in the United States as COMSAT retained its monopoly position with the sole right of access to INTELSAT.[52]

In an attempt to increase its efficiency, in 1995 INTELSAT initiated a review and subsequent revision of the internal mechanisms of two of its major service lines – lease and channel/carrier – to make the organisation more responsive to its customers. Since staff handling channel/carrier orders, known as booking, had been assigned to specific satellites, a customer with multiple services had to work with several staff members, thus increasing the likelihood of errors, whose resolution in turn required customers to negotiate back and forth between several staff.[53] In response to these problems, a team was formed, led by non-booking staff, which redesigned the booking process and simplified the bureaucratic maze for customers. Appreciating the difference in values held by its multicultural staff, INTELSAT made new customer/region assignments based on staff business culture and/or language when possible. These regional assignments, in conjunction with booking staff providing customer consulting services, significantly reduced throughput time, with 98 per cent of new service transactions processed in one to three days, as compared with the previous

---

52 In the Matter of Direct Access to the INTELSAT System, 14 FCC Rcd 15703 (1999).

53 Carole Congram, Peggy Slye, and Priscilla Glidden, "Transformation at INTELSAT: sometimes the tortoise beats the hare," *Team Performance Management: an International Journal* 5, no. 6 (1999): 196.

15 to 20 days for most such transactions, and billing errors became virtually non-existent.[54] In regards to lease services, a group was formed in May 1996 to redress the lease process, since a lease is customized to customer specifications, and to distinguish its role from sales. As a consequence of changes made to lease procedures, including incorporating the lease staff as members of their respective regional sales team, the productivity of both the sales staff and lease staff increased.[55] These efforts at reform demonstrate a commitment to the conviction that management, as opposed to ownership, is the key to efficiency. Lastly, in April 1997 INTELSAT's Assembly of Parties affirmed a Board of Governors decision to eliminate, except for technical interference, the anticompetitive Article XIV(d) of its Charter.[56]

Despite INTELSAT's effort to increase its customer base through direct access and to improve its competitive position, there remained the lack of true competitive access to and ownership of INTELSAT at the national level as well as the absence of a fully competitive charging structure that was responsive to the market and correlated to geography, volume and diverse telecommunications sectors.[57] In lieu of these enduring quandaries, British Telecom had suggested in the early 1990s that INTELSAT be fully privatised, which was followed by a similar proposal by COMSAT in 1993. In the face of substantial resistance to these proposals, the INTELSAT Executive recommended to the Assembly of Parties in 1994 that a Working Party be formed to advise and implement competitive reform.[58] In August 1995 the Assembly of Parties convened to consider the Working Party's recommendation to create a commercial subsidiary, which led, among other things, to the US Clinton Administration and COMSAT proposal in February 1996 to restructure INTELSAT into two groups, one being a continuance of the intergovernmental satellite operator committed to maintaining INTELSAT's core services under the existing INTELSAT Agreement; and the other, a separately chartered private affiliate to offer broadcast and private network services, in which the current Signatories' ownership

---

[54] Ibid, 200.

[55] Ibid, 196 and 198.

[56] INTELSAT, Assembly of Parties, *Record of Decisions of the Twenty-First Meeting*, AP-21-3E, at paras. 10–12 (14–16 April 1997).

[57] Pelton and Edelson, "Reinventing INTELSAT and INMARSAT," 32.

[58] INTELSAT, Assembly of Parties, *Record of Decisions of the Nineteenth Meeting*, AP-19-3E, at para. 24 (25–28 October 1997).

would be rapidly and substantially diluted, by up to 80 per cent spread over two tranches, by the offering of shares to the public, and for which treaty-granted protections would be abolished, and each entity would be assigned assets aimed at providing switched and non-switched business respectively, but with no restrictions prohibiting competitive entry into any service.[59] Although this proposal with its draconian ownership changes was overwhelmingly rejected by the INTELSAT Assembly of Parties in 1997,[60] it ultimately did lead to the creation of New Skies Satellites, N.V., based in the Netherlands, as a spin-off company to which five satellites in orbit and one under construction were transferred in late 1998.

In 1999, in accord with the WTO General Agreement on Trade in Services (GATS), the FCC granted US earth station operators limited three-year authorisations to operate with New Skies in the US market, conditional on the company taking steps to become independent of INTELSAT, including an Initial Public Offering (IPO) to dilute the ownership of INTELSAT signatories, which was successfully concluded in October 2000.[61] Indeed, the final piece of the puzzle was the move to the full privatisation of INTELSAT, which began in earnest in 1999, shortly after the creation of New Skies. It culminated with decisions taken at the Twenty-fifth Meeting of the INTELSAT Assembly of Parties in November 2000 that effectively implemented the privatisation decision, predicated upon substantial amendment of the INTELSAT Agreement and including specific measures such as the transfer of the satellite assets to a separate company; the selection of Bermuda as the jurisdiction of incorporation for the privatised company; the selection of the United States and United Kingdom as the two Notifying Administrations, with the satellite network filings for all satellites then in orbit transferred to the United States;[62] and implementation of a number

59 Contribution of the Party and Signatory of the United States, *INTELSAT Future Structure*, IWP-3-5E (16 February 1996).

60 INTELSAT, Assembly of Parties, *Summary Minutes of Discussions at the Twenty-First Meeting*, AP-21-4E, at paras. 63-301 (14–16 April 1997).

61 The ORBIT Act: An Examination of Progress Made in Privatising the Satellite Communications Marketplace: Hearing Before the Subcommission on Telecommunications and the Internet Commission on Energy and Commerce, 109th Cong. 5 (2005) (statement of Donald Abelson, Chief, International Bureau, Federal Communications Commission).

62 The effect of this was to significantly strengthen the US position in the global marketplace, as INTELSAT's 22 satellites and accompanying optimal orbital locations were transferred under applicable ITU procedures to the United States.

of pro-competitive measures relating to the governance and ownership of the privatised company including a commitment to a future dilution of the ownership interests of the former Signatories.

This latter stipulation is in accord with the property rights school which suggests that privatisation in the form of a change in ownership, an alteration of the structures of property rights, will improve the incentives for productive efficient performance.[63] The property rights school recommends the privatisation of public enterprises operating in competitive markets, for competition compels companies to improve their performance, that is, increase allocative and productive efficiency in order to profit. It would thus seem that an improvement in the economic performance of the public enterprise sector is more likely to result from an increase in market competition than from a change in ownership.[64]

In 1999, the US Congress chartered the Commission to Assess U.S. National Security Space Management and Organisation to evaluate US national security space and ways in which it could be strengthened.[65] According to the White House, as stated in *A National Security Strategy of Engagement and Enlargement*, "We are committed to maintaining U.S. leadership in space. Unimpeded access to and use of space is a vital national interest – essential for protecting U.S. national security, promoting our prosperity and ensuring our well-being. ... We will maintain our technological superiority in space systems, and sustain a robust U.S. space industry and a strong, forward-looking research base."[66] Such views were, arguably, incorporated in the Open-market Reorganisation for the Betterment of International Telecommunications (ORBIT) Act, which was agreed to and passed by both the US House and Senate in the 106th Congress on 24 January 2000, though the legislation was initially proposed in 1997 in concurrence with the WTO General Agreement in Trade in Services.

---

63 Colin Kirkpatrick, "Privatisation in Less Developed Countries: An Overview," in *Privatisation in Less Developed Countries*, eds. P. Cook and C. Kirkpatrick (Brighton: Wheatsheaf Books Ltd., 1988): 19.

64 Ibid, 31.

65 Haller and Sakazaki, *Commercial Space and United States National Security*, 6.

66 White House, A *National Security Strategy of Engagement and Enlargement*, Part II (December 1999): 12.

## 4. *The Role of the Orbit Act – Pre- and Post-Privatisation*

### 4.1. *The ORBIT Act and the Role of Lockheed Martin Corporation*

In June 1997 House Commerce Committee Chairman Thomas J. Bliley (R-VA) introduced in the 105th Congress Bill H.R. 1872 to overhaul the Communications Satellite Act of 1962 to promote competition and privatisation in satellite communications by requiring the privatisation of INTELSAT and INMARSAT. In support of Bliley's Bill, on 24 March 1998, at a hearing on International Satellite Reform before the Subcommittee on Communication of the US Senate, John Sponyoe, the Chief Executive Officer of Lockheed Martin Global Telecommunications, presented a statement in which he described a strategic partnership with COMSAT, which, it should be noted, was at a time when the telecommunications market was thriving and military sales were flat. According to Sponyoe, Lockheed Martin, one of the largest defence contractors in the United States, had created a wholly owned subsidiary, Lockheed Martin Global Telecommunications (LMGT), in an endeavour to focus its business efforts in the commercial telecommunications services market and become a leading provider of global telecommunications services, and, upon the creation of LMGT, had announced its intention to acquire COMSAT.[67] "The rationale for this acquisition is very straightforward – to marry our own technological and entrepreneurial assets with Comsat's 37 years of experience as a provider of satellite services."[68] Lockheed Martin, in essence, proposed a business plan that required an enabling Act of Congress, since the Satellite Act of 1962 prohibited greater than 50 per cent aggregate common carrier ownership in COMSAT.

In May of 1998, Bliley's Bill H.R. 1872 passed as amended the House of Representatives and was then referred to the Senate committee, though it was not addressed by the Senate during the 105th Congress and thus failed to be enacted as law, which was a setback to Lockheed Martin's stated object of acquiring COMSAT. According to Sponyoe, Lockheed Martin was "poised to make a 2.7 billion investment in COMSAT. ... [and was] not

---

67 Hearing on International Satellite Reform, Before the Subcommittee on Communications, Committee of Commerce, Science, and Transportation, United States Senate, 3 (1998) (Statement of John Sponyoe, Chief Executive Officer, Lockheed Martin Global Telecommunications). [Hereinafter Sponyoe Statement].

68 Ibid.

pursuing this investment for the purpose of preserving the status quo at COMSAT – far from it. We want to buy COMSAT, transform it into a normal US commercial business operation, and integrate it into LMGT to form a strong new competitive entrant in the global telecommunications services marketplace."[69] In addressing the fact that Lockheed Martin's acquisition of COMSAT would give it a significant interest in INTELSAT, Sponyoe stated that Lockheed Martin had "no intention of buying COMSAT to acquire an 18% share either in a 'mini-United Nations' or of a diminishing asset. … [W]hatever perceived advantages INTELSAT may or may not have in its current incarnation, these advantages are certainly not reflected in its steadily decreasing market share. … Indeed, INTELSAT's current position in the US-international market *vis-à-vis* other satellite and terrestrial competitors is so far from anything that could be accurately termed 'dominant' that I have to wonder whether its current structure might not pose a greater threat to itself than to its competitors. The INTELSAT Lockheed Martin wants to be part of is one that can soon be a viable commercial system that operates in a manner indistinguishable from any other commercial system. … This is why we see our combination with COMSAT as a means for achieving not only own business objectives, but major US policy objectives as well."[70]

Diane Hinson, Vice President and General Counsel for INTELSAT, said in an interview on 4 May 1998, that INTELSAT Signatories were "a little baffled … at how the United States Congress purports to pass legislation that governs them."[71] Some INTELSAT officials, who supported the Bill's goals, opposed the unilateral dictation of which services INTELSAT could provide to US customers, since H.R. 1872 proposed to restrict the ability of INTELSAT and INMARSAT to provide lucrative new satellite services, such as high speed Internet access or direct-to-home TV broadcasting in the United States, if they failed to privatise.[72] Although privatisation and reform of the telecommunications sector had diffused rapidly internationally in the 1990s, many states had yet to fully engage in this policy diffusion process, and PTT Signatories to INTELSAT were loathe to allow competition to reduce the prices they charged consumers.[73] Sponyoe, however,

69 Ibid, 3–4.

70 Ibid, 4.

71 Juliana Gruenwald, "Satellite Organisation Pins Hopes on Senate to Stop Privatisation Bill," *Congressional Quarterly Weekly* 56, no.19 (1998): 1236.

72 Ibid.

73 Michael Lynch, "Space Balls," *Reason* 30, no. 1 (1998): 14–15.

contended that "as a result of market incentives and the persistent market opening efforts of the US government over many years, there is an *irreversible* trend toward market liberalisation and privatisation around the world."[74] Sponyoe further charged that "the marketplace imperatives that compel this transformation are well understood by INTELSAT management and its *leading* Signatories."[75] Indeed, due to pressure from both within and without the organisation, the 24th INTELSAT Assembly of Parties in 1999 resolved to transform INTELSAT from a public intergovernmental treaty organisation into a private corporation. In light of this resolution, the US Department of State and some legislators warned that unilateral US legislation mandating INTELSAT privatisation would now be unnecessary and potentially counterproductive.[76]

Such views may account in part for that the fact that it took eighteen months, until September of 1999, for Lockheed Martin, frustrated by months of delay in securing the required regulatory approvals to complete the first phase of the merger, to finally gain FCC authorisation to acquire, through its wholly owned subsidiary, Regulus, LLC, COMSAT Government Services, a COMSAT subsidiary, and to purchase up to 49 per cent of COMSAT stock. Following these initial FCC regulatory approvals and COMSAT shareholder approval, the first phase of the transaction was completed, which entailed a tender offer for 49 per cent of the outstanding shares of COMSAT common stock that was valued at approximately US$1.2 billion.[77] President and CEO of COMSAT, Betty Alewine, hailed the deal as a 'strong pro-competitive merger' that would increase global telecommunications competition, since Lockheed Martin's space assets, including rocket and satellite building capabilities, in combination with COMSAT's satellite services network, would strengthen Lockheed Martin's position as an industry power and thereby increase its ability to compete with The Boeing Company.[78]

---

[74] Sponyoe Statement, 9; emphasis added.

[75] Ibid, 6; emphasis added.

[76] Katkin, "Universal Global Interconnection After INTELSAT," 17.

[77] "Lockheed Martin Completes Acquisition Of Comsat Corporation," *SPACEandTECH Digest*. Andrews Space & Technology (2000). Available: http://www.spaceandtech.com/digest/sd2000-21/sd2000-21-005.shtml.

[78] n.a., Congress Keeps Lockheed Martin-Comsat Merger On Hold Until Legislation Passes – Brief Article (1999). Available: http://findarticles.com/p/articles/mi_m3457/is_20_11/ai_55938901.

The goal of the ORBIT Act,[79] to promote competition in the international satellite communications market by encouraging the timely and procompetitive privatisation of INTELSAT and INMARSAT, was strongly supported by the Clinton Administration. Thus, on 17 March 2000, the ORBIT Act,[80] which set forth specific criteria relating to the privatisation of INTELSAT and INMARSAT, was enacted by Congress and signed by President Clinton. Then, following FCC authorisation on 31 July 2000, COMSAT assigned its licenses and authorisations to Regulus, LLC, thereby enabling Lockheed Martin's successful purchase of the remaining shares of the COMSAT Corporation in a stock swap valued at US$790 million, as the second phase of the transaction, the one-for-one tax-free exchange of Lockheed Martin common stock for COMSAT common stock, was completed, and COMSAT stock ceased to be traded on 3 August 2000.[81] Owing to the lengthy approval process, however, COMSAT shareholders lost an estimated US$300 million in purchase value because of a drop in Lockheed Martin stock, allegedly due to the company's two-year toil to overcome complex congressional and regulatory barriers, but also attributable to other problems within Lockheed Martin.[82] A series of launch failures in 1999 destroyed approximately US$4 billion in rockets and payloads, which led to an inquiry that found poor management practices and quality-control problems that prompted the company to restructure and sell many non-core operations, since its profits were two-thirds lower in 1999 than the prior year.[83] In 2000, the Pentagon reportedly bailed out Lockheed Martin by agreeing to buy 24 C-130J transports and awarding it other Pentagon contracts.[84] Despite its internal difficulties, Lockheed Martin's perseverance enabled it to finally become the US owner and the largest shareholder in INTELSAT and INMARSAT in August 2000, which had been previously privatised on 15 April 1999, as well as New Skies.

In the aftermath of the attack on the Pentagon and the Twin Towers of the World Trade Center in New York and in light of continued overcapacity in the telecommunications industry and deteriorating market conditions

79 S. 1328, 105th Cong. (1997); H.R. 1872, 105th Cong. (1997).

80 ORBIT Act, Pub. L. 106–180, 114 Stat. 48 (2000).

81 "Lockheed Martin Completes Acquisition of Comsat Corporation."

82 Ibid.

83 n.a., Hoover's Company Profile Database—American Public Companies (2002). Available: http://homepages.wmich.edu/~a2olexse/webresearch.htm.

84 Ibid.

in Central and South America, Lockheed Martin recommitted itself to its core business of US defence, government information technology and homeland security, and thus eliminated the Lockheed Martin Global Telecommunications (LMGT) administrative structure and immediately reassigned certain of the former LMGT businesses and investments to other units of Lockheed Martin while initiating sales of the remaining operations, including its stake in Intelsat, Inmarsat, New Skies and other satellite ventures.[85] Thus, within five months of the privatisation of INTELSAT, in December 2001, the company announced its decision to exit the telecommunications services business, thereby representing a stark reversal of Lockheed Martin's stated commitments in acquiring COMSAT to build a strong new competitive entrant in the global telecommunications marketplace. At the time, Lockheed Martin took a US$1.7 billion charge to write down the COMSAT purchase and, as the company began to dismantle its expensive foray into commercial satellite services, it sought to delay Intelsat, Ltd.'s initial public equity offering, for otherwise it would have to sharply write down its Intelsat investment as it was carried on its books at a much higher value than that implicit in the IPO.[86]

### 4.2. *IPO Delays*

One provision of the ORBIT Act required that the privatised company conduct an IPO of equity securities that would "substantially dilute" the aggregate ownership of the former Signatories of INTELSAT.[87] These entities were either private telecommunications companies or governmental agencies of the countries party to the intergovernmental agreement that formally established INTELSAT, and these entities became shareholders of Intelsat, Ltd. upon its privatisation. On 18 December 2003, Intelsat, Ltd. obtained an extension from the FCC to extend the date by which the company must complete an initial public equity offering from 31 December 2003 to 30 June 2004.

---

85 "Lockheed Martin Quits Telecom Business," *SPACEandTECH Digest*. Andrews Space & Technology (2001). Available: http://www.spaceandtech.com/digest/flash2001/flash2001-106.shtml.

86 Peter de Selding, "IPO Delay Could Gut Revenues From Intelsat's New Fleet," *Space News* (2004). Available: http://www.space.com/spacenews/archive04/ipoarch_051004.html.

87 ORBIT Act, S.376. 106 Cong. § 621(2) (2000).

In February 2004, the company announced its intention to conduct an IPO of its ordinary shares in an amount of up to US$500 million, but then withdrew its planned IPO when an amendment to the ORBIT Act was signed into law on 18 May 2004, changing the date by which Intelsat, Ltd. needed to conduct an IPO to as late as 31 December 2005, thus giving the company the opportunity to explore other options, for which it retained Merrill Lynch and Company and Morgan Stanley.[88] Senator Conrad R. Burns (R-Montana) introduced the legislation that permitted Intelsat, Ltd. to delay its IPO by up to 18 months because several US investors in Intelsat, notably Lockheed Martin as the largest single shareholder with 24 per cent of shares, stood to lose millions of dollars since the market conditions were not conducive for conducting a successful IPO.[89] The quandary Intelsat faced was that its near billion dollar purchase in March of 2004 of six North American satellites and related customer service contracts of Loral Space and Communications Corporation included services, such as direct-to-home TV broadcasting and high speed Internet access, that Intelsat was forbidden to offer in the United States before its IPO was completed under the terms of the ORBIT Act. Although the FCC granted Intelsat a six-month grace period to continue offering such services so as not to interfere with the Loral purchase, the acquisition placed pressure on Intelsat to complete an IPO as its projected revenues from the acquisition would be jeopardized.

### 4.3. *Zeus – Private Equity Acquisition of Intelsat*

Another amendment to the ORBIT Act was then signed into law on 25 October 2004, providing an alternative method by which Intelsat could comply with the privatisation requirements of the Act, that is, the dilution of share ownership, by adding new subsections (F) and (G) to Section 621(5). Thus, on 22 December 2004, the FCC issued an Order granting applications filed by Intelsat and Zeus Holdings Limited (Zeus), a company

---

88 Intelsat, Ltd., "Intelsat Announces Decision to Withdraw Planned Initial Public Offering of Shares and Intention to Explore Strategic Alternatives," (21 May 2004). Available: http://www.iintelsat.com/aboutus/press/release_details.aspx?year=2004&art=20040521_01_EN.xml&lang=en&footer=61.

89 Peter de Selding, "IPO Delay Could Gut Revenues From Intelsat's New Fleet," *Space News* (10 May 2004). Available: http://www.space.com/spacenews/archive04/ipoarch_051004.html.

formed by a consortium of private equity funds advised by Apax Partners, Apollo Management, Madison Dearborn Partners and Permira, to transfer control of certain FCC authorisations from Intelsat to Zeus, as the Commission concluded that approval of the applications would serve the "public interest, convenience, and necessity," the nebulous terms employed by the Communications Act of 1934 and retained in subsequent legislation.[90] On the following day, 23 December 2004, in conjunction with this Order, Intelsat filed a Petition for Declaratory Ruling and attached certification, pursuant to Section 621(5)(F) of the ORBIT Act, that upon consummation of the transaction with Zeus, Intelsat would be in compliance with the said requirement as it would achieve substantial dilution of the aggregate amount of former signatory financial interest in Intelsat, which the FCC granted on 8 April 2005.[91] On 28 January 2005, Intelsat announced the successful closing of the amalgamation under Bermuda law of Intelsat and a subsidiary of Zeus, with the result that those individuals and organisations that held shares in Intelsat immediately prior to the closing of the transaction were entitled to receive US$18.75 in exchange for each share.[92] Hence, the former intergovernmental treaty based satellite organisation became a private equity acquisition, arguably to ensure the financial wellbeing of its largest shareholder, Lockheed Martin.

Zeus purchased Intelsat for about US$3 billion, putting up US$515 million and borrowing the rest, and then during 2005, the group paid itself two dividends, one being about US$340 million, which the company had to borrow, and the other being about US$198 million, which was financed from the company's free cash flow.[93] The new private equity owners also paid themselves about US$70 million in management fees, which resulted in the four firms comprising Zeus essentially recouping all the cash they had invested in Intelsat while retaining virtually all of the stock by the end of 2005. Within a period of two years, the private equity owners drained much of Intelsat's equity value as they engaged in a series of acquisitions which increased the company's debt to more than US$11 billion.[94]

---

90 Abelson, "Written Statement of Donald Abelson, Chief, International Bureau," 3.

91 In the Matter of Intelsat, Ltd., 20 FCC Rcd 8604 (2005).

92 "Intelsat Announces Completion of Acquisition by Zeus Holdings Limited," Intelsat, Ltd. Available: http://www.intelsat.com/aboutus/press/release_details.aspx?year=2005&art=20050128_01.

93 Steven Pearlstein, "Sweet Deals Buried Intelsat in Debt," *The Washington Post*, August 18, 2006: D01.

94 Ibid.

## 4.4. *Mergers and Acquisitions*

With John Sponyoe, former CEO of LMGT, serving as the Chairman of Intelsat's Board of Directors upon its privatisation, Intelsat, Ltd. boarded the merry-go-round of mergers and began to acquire components from, ironically, Lockheed Martin. In an effort to be more market-oriented and assemble a new ground-based infrastructure to complement its global satellite system, in March 2002, Intelsat signed an agreement to acquire the World Systems business unit of Lockheed Martin Corporation, including earth stations located in Clarksburg, Maryland and Paumalu, Hawaii, and also purchased the teleport facilities and related assets of Comsat Digital Teleport, Inc.[95] In building a Global Connectivity Solutions portfolio, Intelsat, while continuing to be a leading provider of satellite communication services offering video, data and voice connectivity, made other acquisitions as well, so that in May 2002, within a year of its privatisation, it began providing end-to-end solutions through its newly established network of teleports, leased fibre and points of presence (POPs) around the globe. This global infrastructure enables Intelsat to earn revenue through the delivery of hybrid services that combine space segment, teleport facilities, cable and other land facilities for end-to-end telecommunications, though the company continues to earn revenue principally by selling satellite transponder capacity and seeks to sustain its leadership position in the fixed satellite services (FSS) sector of the satellite industry.

To further expand its capabilities, in September 2002, Intelsat made an unsuccessful bid to acquire the recently privatised or restructured European operator, Eutelsat, then, in May 2004, Intelsat acquired Lockheed Martin's COMSAT General business, which provided satellite-centric telecommunications services and equipment, concentrating on international fixed and mobile satellite systems for clients with quick response and high availability needs, for approximately US$90 million.[96] Intelsat, as noted, also acquired the US domestic satellite fleet operated by Loral Skynet in 2004. Although INTELSAT's privatisation as a means of promoting competition was the stated purpose of the ORBIT Act, in actuality, the industry has since

---

[95] Roy Mark, "Intelsat to Acquire Lockheed Martin Unit," (March 18, 2002). Available: http://www.internetnews.com/bus-news/article.php/992961.

[96] "Lockheed's COMSAT General to be Acquired by Intelsat," Lockheed Martin Corporation. Available: http://www.defenseaerospace.com/cgiin/client/modele.pl?prod=38658&session=dae.21650007. 1152057590.RKsA9sOa9dUA.

witnessed increased consolidation, with, of course, the approval of the FCC, as exemplified by Intelsat's acquisition of PanAmSat in July 2006 for US$6.4 billion, including US$3.2 billion assumed debt, creating the largest global satellite company with a fleet of 51 satellites.

## 5. *ITSO – The Residual Intergovernmental Organisation*

The state of indebtedness resulting from this merger was of serious concern to ITSO, the residual intergovernmental organisation that monitors Intelsat's compliance in providing international public telecommunications services in conjunction with adherence to three Core Principles, namely, as stated in the ITSO Agreement, "to maintain global connectivity and global coverage; serve its lifeline connectivity customers, and provide non-discriminatory access to the Intelsat system."[97] ITSO must thus ensure that Intelsat maintain the technical interconnection capability to carry communications to and from any country or territory within and between the five regions, as defined by the International Telecommunication Union (ITU), of Africa, the Americas, Asia and the Pacific, Europe, and the Middle East. ITSO must also assure that all users and prospective users enjoy fair and equal opportunity to access the Intelsat system. Finally, ITSO must ensure Intelsat's adherence to its Lifeline Connectivity Obligations (LCO) customers, which includes those countries that are either low income, as classified by the World Bank, or have a low teledensity, as defined by the ITU. At the Intelsat 25th Assembly of Parties in November 2000, 69 countries qualified for LCO protection under the income or teledensity eligibility criteria, and 41 additional countries or territories were identified as qualifying for "Petition Eligible" LCO Protection for either all or part of their international traffic. These countries then had until 1 August 2000, to petition the Assembly of Parties for LCO protection on the basis that either there was no cost effective alternative provider of a service equivalent to Intelsat's services or, among other criteria, were facing emergency conditions, such as an earthquake or war.[98] Also, new countries, created after 18 July 2001, were deemed eligible to join ITSO and apply for LCO protection.

---

97 "ITSO Mission & Role," ITSO. Available: http://www.itso.int/php_docs/tp11_itso.php?dc=mission.

98 Katkin, "Universal Global Interconnection After INTELSAT," 26–27.

ITSO has questioned Intelsat's commitment to the ITSO Agreement as the company has pared down its public service obligations and decreased satellite coverage in several sensitive regions, including Asia and Africa.[99] ITSO contended that the increase in Intelsat's debt level due to the purchase of PanAmSat, to which it objected, but had no authority to prevent, made the company unstable and could thus trigger a bankruptcy filing or a creditor-forced sale of assets, and, as there were no legal covenants preventing Intelsat from divesting any of its satellites, the new owner would not be bound to adhere to the Core Principles established by the ITSO Agreement. In seeking conditions on the PanAmSat merger, ITSO thus petitioned the FCC to modify the space station licenses held by Intelsat for the use of certain orbital locations and associated radio frequency assignments to ensure that Intelsat and any successor satellite operator using the orbital locations and associated frequencies would be bound by the ITSO Agreement, essentially forcing Intelsat to assure that some of its assets would continue to be available for public-service use, even if owned by another company, in the event of a bankruptcy. ITSO also asked that Intelsat place a lien, a letter of credit or some other binding guarantee on at least five satellites and orbital slots that could together continue to offer global coverage, but the FCC refused to consider the issue during the regulatory review of the proposed merger with PanAmSat, and suggested that ITSO's concerns could be more properly addressed in a separate proceeding initiated under Section 316 of the Communications Act.[100] ITSO subsequently initiated this proceeding,[101] and ultimately the FCC agreed to impose licensing conditions on the Intelsat satellites that would ensure that Intelsat and any successor entity adhere to the requirements of the Public Services Agreement (PSA) entered into between Intelsat and ITSO, although it declined to impose any financial conditions in the form of a lien or similar instrument.[102]

---

99 Comments of the International Telecommunications Satellite Organisation (ITSO) (filed 14 November 2004), submitted in Constellation, LLC et al., Consolidated Application for Authority to Transfer Control of PanAmSat Licensee Corp. and PanAmSat H-2 Licensee Corp., IB Docket No. 05-2.

100 Ibid.

101 Petition of the International Telecommunications Satellite Organisation (ITSO), IB Docket No. 06-137 (filed 10 July 2006).

102 Petition of the International Telecommunications Satellite Organisation (ITSO), IB Docket No. 06-137, Order of Modification, 23 FCC Rcd 2764 (2008).

While ITSO's more limited role, straddling as it is between the needs of developing states and the private sector, is not surprising in this era of unprecedented corporate enclosure of the commons, from water resources and forests to biodiversity and knowledge, it is nonetheless the case that a privatised commons, in this case, space, is no longer a commons, it is private property. The Outer Space Treaty, formally the Treaty on Principles Governing the Activities of States in the Exploration and Use of Outer Space, Including the Moon and Other Celestial Bodies, which represents the basic legal framework of international space law, states in Article I that outer space shall be used for the benefit and in the interest of all countries and shall be in the province of all mankind.[103] At their core, though, policies emanating from hegemonic nation states and arguably undemocratic global governing bodies of the World Bank, the International Monetary Fund (IMF), and the WTO, are reflective of the shift from social values, as represented by the Outer Space Treaty, to commercial values, as evident in the private equity ownership of Intelsat. The increasing levels of interdependence and vulnerability enhanced by technological change necessitate innovative, democratic forms of global political authority and governance, especially in regard to the commons, but precedent is set by private corporate control and influence in the multilateral institutions forming the global trade architecture whose strategies are facilitated by individual ascendant states, such as the United States.

For example, the former treaty-based INTELSAT was immune from US trade policies, but the private, US-licensed Intelsat, Ltd. is subject to such policies, thus, in the event that the provision of service to a lifeline customer, that is, a country dependent on Intelsat for carriage of their intercontinental telecommunications traffic, as in accord with the ITSO Agreement, should conflict with a subsequently enacted US policy or law, the US law will supersede Intelsat's prior obligations.[104] In essence, lifeline countries, which include countries not allied with the United States, such as Afghanistan, Sudan, North Korea, Somalia, and Cuba, could be cut off from the global telecommunications network by the imposition of any future US economic trade sanctions. Although one might reason that the

103 Art. I, Treaty on Principles Governing the Activities of Sates in the Exploration and Use of Outer Space, Including the Moon and Other Celestial Bodies, entered into force 10 Oct. 1967, 610 U.N.T.S. 205, 18 U.S.T. 2410. Available: http://www.state.gov/t/ac/trt/5181.htm.

104 Katkin, "Universal Global Interconnection After INTELSAT," 36.

United States would forbear from such actions to minimize negative diplomatic ramifications, in 2002 the United States blocked deployment of a planned Ku-band Intelsat satellite (Intelsat APR-3), capable of providing strategic landmass coverage of China, Russia, India, and the Middle East, and to which SINOSAT, a commercial telecommunications agency of the Chinese government, had purchased the right to use six transponders. The Intelsat APR-3 satellite was to have been launched by the China Great Wall Industry Corporation, a Chinese government agency, but the US State Department denied Intelsat's application for an export license that would have permitted the launch, though it did not violate any US law or policy, and Intelsat, thus, terminated its deployment, providing evidence that privatisation does not eliminate politicization.[105]

## 6. *BC Partners/Serafina – From Private Equity to Private Equity*

The credit-rating service Moody's said on 8 January 2007 that Intelsat's refinancing transactions would not change Intelsat's continued high level of debt, thus leaving Intelsat's debt rating unchanged, as Moody's contended that, given Intelsat's private equity investor shareholder mix, the company was more likely to pay dividends than use its cash flow for debt reduction.[106] Indeed, Intelsat decided to increase its capital spending plan for 2007, seeing more reasons to focus on investment than for reducing its debt, which was arguably a good decision for the company's investors as the debt did not deter the bidding for Intelsat when the company placed itself up for auction.

Although the practice of a private equity firm buying a company from another private equity firm, referred to as a secondary buyout, is still relatively rare, on June 20, 2007, Intelsat announced the signing of a definitive agreement for the purchase of 76 per cent of Intelsat, Ltd. by BC Partners, an international private equity firm based in the United Kingdom but organised under the laws of Guernsey. BC Partners, which was formed in 1986, operates through integrated teams based in Geneva, Hamburg,

---

105 Ibid, 36,40, and 41.

106 "Moody's assigns ratings to Intelsat Bermuda's Proposed Financing," Moody's Investor Service. Available: http://www.moodys.com/research/Moodys-assigns-ratings-to-Intelsat-Bermudas-Proposed-Financing–PR_125775.

London, Milan, Paris, and New York. In the transaction valuing Intelsat's equity at approximately US$5.03 billion, BC Partners received US$5.11 billion in financing commitments from Credit Suisse, which advised Intelsat during the transaction, Banc of America Securities, and Morgan Stanley.[107] BC Partners assumed Intelsat's debt, which, just prior to the transaction, as of 31 March 2007, stood at approximately US$11.4 billion, and an additional US$3.85 billion in debt as a result of the anticipated financings at closing, thus the enterprise valuation implied by the transaction was approximately US$16.4 billion.

To facilitate the transfer of control of Intelsat, Serafina Holdings Limited was formed in Bermuda and was effectively controlled by BC Partners, as it held approximately 71 per cent of the equity and voting interests in Serafina through 41 subsidiary investment funds (the "BCP Funds"). Two funds controlled by Silver Lake Group, L.L.C., a US-based investment firm, collectively held approximately 16.84 per cent of the equity interests in Serafina. Other investors in Serafina included Banc of America Capital Investors, V, L.P., CSFB Strategic Partners III, L.P., which was indirectly controlled by Credit Suisse, and 13 members of Intelsat's management team. Since virtually all of the foreign investment in Serafina came from entities or individuals whose home markets were WTO Member states, the FCC, in accord with US WTO commitments, presumed that such investment would serve the US "public interest, convenience, and necessity," and thus approved Serafina's proposal to acquire Intelsat through its wholly-owned subsidiary, Serafina Acquisition Limited, also a Bermuda company. Upon summation of Serafina's acquisition of all of the equity and voting interests in Intelsat, Intelsat Bermuda, an indirect, wholly-owned subsidiary of Intelsat Holdings, Ltd., transferred substantially all of its assets and liabilities to Intelsat Jackson Holdings Ltd., a new wholly-owned subsidiary of Intelsat Bermuda, including all of the existing indebtedness of Intelsat Bermuda, and the debt that ensues from the acquisition of Intelsat by Serafina was assigned (by contract, merger or otherwise) to Intelsat Bermuda.[108]

---

107 "BC Partners to Acquire Majority of Intelsat," Intelsat, Ltd., (19 June 2007). Available: http://www.spacered.com/newa/viewpr.rss.spacewire.html?pid=22894.

108 Intelsat Holdings, Ltd. and Serafina Holdings, Ltd., *Application for Satellite Space and Earth Station Authorisations for Transfer of Control or Assignment, Approved by OMB 3060-0678*, Filed 10 August 2007.

## 7. *Developments Subsequent to the Acquisition by BC Partners*

Although private equity firms often acquire a company to break it up into smaller units and thereby, not only recouping their initial investment, but earning an increased amount of capital, Intelsat has remained intact and BC Partners have retained Intelsat's existing management team and operational staff, including the Chief Executive Officer, Dave McGlade, who joined Intelsat in April 2005. Intelsat, as the world's leading provider of FSS worldwide, serving the media, network services and government customer sectors, is, according to its own accounts, considered the gold standard in satellite operations, with the hallmark of the system being its transponder reliability, with a 99.997 transponder availability rate since 1985.[109] Maintaining this record, coupled with continued upgrading of its fleet, will undoubtedly help position the company and its current private equity owners for a future transaction that would enable the current owners to cash out their investment.

To the extent that an exit strategy looms, it would now appear to take the form of an IPO, rather than a sale of the company to either another satellite operator or to another strategic investor. Intelsat made a preliminary filing with the FCC in November 2011, seeking FCC approval for a transfer of control that could potentially occur should a contemplated IPO result in a sufficient dilution of BC Partners ownership stake that it no longer had control of Intelsat, and the FCC granted this application in May 2012.[110] Shortly thereafter Intelsat filed a Registration Statement (Form F-1) with the U.S. Securities and Exchange Commission looking towards conducting an IPO at some later time.[111]

Undoubtedly, the financial goal of preparing a profitable exit from the business affects Intelsat's strategies for growth. The company locked in interest rates for its debt contracts in 2008, with respect to the next large

---

109 Intelsat, Ltd., Resources, (2003a). Available: http://www.intelsat.com/resources/satellites.aspx.

110 Intelsat Global Holdings, S.A, Applications to Transfer Control of Intelsat Licenses and Authorisations from BC Partners Holdings Limited to Public Ownership, 27 FCC Rcd 5226 (2012).

111 Intelsat Global Holdings SA, Form F-1 Registration Statement Under the Securities Act of 1933, Filed: 18 May 2012.

bond payment that was due in 2010.[112] Subsequently, in 2011, the company refinanced through new borrowings and redemption of prior debt issues approximately $5.9 billion in debt, pushing the maturity of a significant portion of its debt to 2016 or beyond.[113]

### 7.1. *Joint Ventures and Long Term Contracts*

In an effort to secure its position in the FSS market, Intelsat is increasingly moving to long-term contracts and new joint ventures, as exemplified by its joint venture with a South African investor group. Led by Convergence Partners, the South African investor group utilised project financing to build and launch a new satellite, called "Intelsat New Dawn," into the 33 degree East orbital location, a position well suited to serve the African continent as the satellite featured a payload optimized to deliver wireless backhaul, broadband and television programming to the continent.[114] Agreements were concluded for financing the project, projected to cost approximately US$250 million. The undertaking was funded with roughly 15 per cent equity and 85 per cent debt, the debt being in the form of non-recourse project financing provided by African institutions. The debt financing was arranged by Nedbank Capital, part of the Nedbank Group, one of South Africa's largest banking groups, and a leading telecommunications project financier in South Africa. Nedbank and the Industrial Development Corporation of South Africa, a self-financing national development finance institution, are the largest participants in the debt funding consortium. In regard to the equity financing, Intelsat is to provide 74.9 per cent and the Convergence Partners-led group 25.1 per cent. Thus, Intelsat's actual cash contribution was roughly on the order of US$25

---

[112] "Intelsat Announces Proposed Offerings of Senior Notes and Repayment of Outstanding Notes Intelsat," Intelsat. Available: http://www.intelsat.com/_files/investors/financial/2008/2008-44.pdf.

[113] Intelsat, Intelsat Announces Tender Offers and Consent Solicitations for Certain Notes of Intelsat Jackson Holdings S.A., Intelsat Intermediate Holding Company S.A. and Intelsat Subsidiary Holding Company S.A. (21 March 2011) Available: http://www.intelsat.com/_files/investors/financial/2011/2011-09.pdf; Intelsat, Intelsat Announces Pricing of Senior Notes (22 March 2012) Available: http://www.intelsat.com/_files/investors/financial/2011/2011-10.pdf.

[114] "Intelsat, Convergence Partners Form New Dawn Joint Venture to Grow Africa's Telecommunications Infrastructure," Intelsat. Available: http://www.intelsat.com/press/news-releases/2008/20081209.asp.

million.[115] Such creative financing, in this case in the form of a joint venture, enabled Intelsat, notwithstanding its heavily indebtedness, to secure new satellites without itself bearing the financing. Unfortunately, following the launch of the New Dawn satellite in 2011, the C-band payload on the satellite was rendered inoperable due to an in-orbit anomaly, significantly reducing the overall functionality of the satellite and resulting in an insurance claim of approximately US$118M.[116]

Long term contracts perform a similar function for Intelsat, as they too enhance the company's long term security. In this regard, Intelsat has agreed on a multi-year contract with GlobeCast, which has invested in expanded occasional use capacity spanning Intelsat's entire satellite fleet. Intelsat has dedicated approximately 140 transponders across its fleet to meet GlobeCast's growing demand for high definition services, live sports broadcasting and breaking news coverage.[117]

Similarly, Intelsat's contract with the Australian Defense Force (ADF), which agreed to purchase a specialized UHF communications payload aboard an Intelsat satellite which was successfully launched in 2012, is illustrative of the effort undertaken by the company to diversify the ways in which it finances new satellites, lessening the costs borne by Intelsat. The contract with the ADF is designed to provide a UHF-band mobile communications payload aboard the IS-22 satellite, which was deployed over the Indian Ocean and whose primary mission is to supply C- and Ku-band communications for maritime users and telecommunications services between Asia and Africa.[118] As part of the "hosted payload" contract valued at approximately US$167 million, Intelsat arranged for the construction and integration of the UHF payload with its satellite, and will operate the ADF payload and provide related services for 15 years after the launch.[119] The UHF payload, which is compliant with US Department of Defense Mil-Std-188-181 and Volna Treaty (Russian) requirements for interoperability, will provide Australian ships, aircraft and troops with an advanced global

115 Ibid.

116 "Intelsat Announces Fourth Quarter and Full Year 2011 Results," Intelsat. Available: http://www.intelsat.com/_files/investors/financial/2011/2011-4Qer.pdf.

117 Intelsat Leases Fleet-wide Capacity to GlobeCast, *Satellite Today* (10 September 2009).

118 "Australian Defence Force Contracts for Hosted Payload on New Intelsat Satellite," Intelsat. Available: http://www.intelsat.com/press/news-releases/2009/20090427-1.asp.

119 Ibid.

communications capability.[120] Intelsat CEO David McGlade has indicated that the company's contract with ADF exemplifies how the United States and other military forces should systematically employ hosted payloads aboard otherwise commercial satellites.

Hosted payloads would certainly enhance Intelsat's financial position as the company develops its strategy for long-term growth by creating new revenue streams on satellites that are designed to optimize capacity. In an effort to replace profitable satellites nearing retirement through partnerships that minimize Intelsat's capital investment, the company is likely to pursue other agreements or contracts similar to that conducted with the ADF and the South African investor group led by Convergence Partners, as well to provide payload space for military and other government agencies aboard commercial telecommunications satellites.

### 7.2. *Intelsat and the US Military: Issues and Opportunities*

Intelsat has pursued a stronger relationship and ties with the US Department of Defense as well as other US government users. In this regard, in 2003, it created the subsidiary, Intelsat Government Solutions Corporation, to address the communications needs of key prime contractors serving the United States as well as NATO government users. Then, in November 2004, after it completed the acquisition of assets, contracts, and staff of the COMSAT General business from Lockheed Martin in late October 2004, Intelsat integrated the two businesses and changed the name of Intelsat Government Solutions Corporation to Intelsat General Corporation. Intelsat General has since consistently sought to provide vital tactical communications capacity to the US military and its allies. To further its efforts, Intelsat General has formed alliances to expand its offerings. For example, Paradigm Secure Communications (Paradigm), a wholly owned subsidiary of the European Aeronautic Defence and Space Company (EADS), which provides satellite communication services to the UK Ministry of Defence as it was specifically created to service military, defence and government organisations with specialized military satellite communications, has expanded its distribution partnership with Intelsat. Intelsat General signed a contract in 2008 with DRS Technologies to provide the US military with X-band satellite capacity on the Skynet satellites of Paradigm,

120 Ibid.

and, later, a contract to provide UHF services on the Skynet satellites to the US Navy.[121]

Due to the increased bandwidth needs of the US State Department, the Department of Defense, and its coalition forces, the US government has come to rely on commercial satellite operators for 80 per cent of its total capacity to meet mission requirements, and more than 90 per cent in battle arenas such as Iraq and Afghanistan. The US government is thus consolidating its bandwidth purchasing programs so as to enable US government agencies, as of 2011, to secure bandwidth directly from satellite operators. To ensure that the US military continues to have access to commercial bandwidth when needed, Intelsat CEO McGlade has posited that the industry needs to have a better sense of the military's future needs.[122] To improve relations between the military and commercial satellite operators, McGlade has stated that there needs to be a better understanding of the common challenges both sides face, a clear demarcation of specific actions that will ensure that commercial services are available when and where needed, as well as new approaches to improve the security of space communications.[123]

### 7.3 *Additional Satellite Acquisitions*

Bolstering US satellite manufacturers, Intelsat signed a contract with Boeing to deliver four telecommunications satellites with an estimated value of US$500 million, thus marking the return of Boeing Satellite Systems to the commercial market, in which of late it has served as more of a spectator than an active participant. The contract with Intelsat is the first order for Boeing's 702B satellite series, an evolution of the 702 satellite, which reportedly provides the flexibility needed to satisfy a wide range of commercial and government satellite requirements as the spacecraft operates in the medium-level power range, with six to 12 kilowatts of on-board power and more flexible, modular design that maintains the spacecraft's advanced technologies.[124] The four Boeing-built geostationary satellites, beginning with the Intelsat 21 and Intelsat 22 satellites, will add new capacity to Intelsat's global satellite fleet as they offer C- and Ku-band

121 Intelsat General Expands UHF, X-band Distribution Partnership with Paradigm, *Satellite Today* (15 September 2009).

122 Lon Rains, McGlade Calls for New DoD Commercial Bandwidth Strategy, *Space News* 19, no. 40 (13 October 2008): 15.

123 Ibid.

124 "Boeing Signs 4-Satellite Contract With Intelsat," Boeing. Available: http://boeing.mediaroom.com/index.php?s=43&item=748.

capacity designed to distribute video, network and voice services across the world regions in which Intelsat operates.[125]

## 8. *Conclusions*

Intelsat, Ltd. clearly faces a host of issues as it seeks to retain its competitive stance, especially as it is encumbered by high debt and private equity owners seeking a profitable exit from the industry. From the perspective of developing countries with inequitable access to communications services, Intelsat seems unlikely to reduce that disparity. The residual intergovernmental organisation, ITSO, is hampered in its regulatory ability to ensure that commercial imperatives don't trump global egalitarianism in a market-driven environment. Although geostationary communications satellites continue to be the primary provider of telecommunications services in areas with low-density populations, particularly rural areas in developing countries, which persist in having few wire line networks because the return on investment is not commercially viable, consolidation in the satellite industry has resulted in fewer players providing services. The admirable goals of the US Communications Satellite Act of 1962 and the commendable aims of the Outer Space Treaty seem to have been lost in the neoliberal environment. The entire process of privatising INTELSAT, from the sale of Intelsat to Zeus Holdings Limited to its ensuing series of acquisitions, was largely undertaken to benefit US corporations, not to create an international satellite fleet better able to meet the telecommunications needs of the economically underprivileged countries of the South. If the wealth and socio-economic health of a state is defined by its ability to participate in the networked economy, these states can be considered impoverished, but ITSO was not endowed with the power to ensure that developing countries' technological dependency and marginalization in this era of accelerated technological change is in fact reduced. Rather, the Lifeline Connectivity Obligations of the privatised Intelsat ended in 2012, insufficient time for developing countries to overcome patterns of dependency. In an era of increased private equity ownership of space assets, the world thus still requires a competitively neutral international transfer mechanism for directly subsidizing economically disadvantaged users in developing states.

---

125 Ibid.

# Chapter Four

## The Evolution of EUTELSAT: A Challenge Successfully Met

*Christian Roisse*

### 1. *Introduction*

This chapter presents an analysis of the European Telecommunication Satellite Organisation (EUTELSAT)[1] before, during and after its restructuring. The term "restructuring" is used, as opposed to that of "privatisation," since privatisation refers to the sale of publicly owned interests to a private entity. The restructuring of EUTELSAT consisted of the transfer of all the assets, operational activities and related liabilities and commitments of an existing intergovernmental organisation (EUTELSAT) to a private entity (Eutelsat S.A.) established in France for this purpose. An intergovernmental organisation was maintained with a transformed structure, role, mission and activities (EUTELSAT IGO). As of today, the legal situation of Eutelsat S.A. is quite specific, and also rather unusual in the arena of both business and international public law, because not only is it operating internationally under national law but it also needs to comply and respect obligations embedded in an amended international treaty safeguarded by the intergovernmental organisation EUTELSAT IGO.

In order to proceed with this analysis, it will be necessary to present how and why EUTELSAT was created (Part 2), how the organisation operated prior to restructuring (Part 3), why it became necessary to restructure

1 In this document "EUTELSAT" refers to the intergovernmental satellite organisation prior to restructuring, "EUTELSAT IGO" refers to the intergovernmental organisation following restructuring and "Eutelsat" refers to the company Eutelsat S.A., established in 2001 at the time of restructuring, together with the holding company it established in 2005, Eutelsat Communications S.A., which is quoted on the Paris Stock Exchange.

(Part 4), how this was done (Part 5) and the results of the restructuring (Parts 6 and 7).

## 2. *Establishment of EUTELSAT*

The post World War II environment led to the creation of the United Nations Organisation. It was in this context that, given the increasing strategic importance of space in general and satellites in particular, in December 1961, the UN General Assembly passed its Resolution 1721 (XVI) Part D, on International Cooperation in the Peaceful Uses of Outer Space.[2] It was expected that "communications by satellite should be available to the Nations of the World as soon as practicable on a global and non-discriminatory basis." In the environment of the time, it was felt that space activities could only be in the hands of governments and should be within the sphere of national sovereignty of States. It was acknowledged that international cooperation was worthwhile to develop space utilization so as to share the risks and the large level of capital and investment expenditure needed for such activities. This co-operation was realised firstly with the creation of INTELSAT[3] and subsequently INMARSAT.[4]

While European states played a critical role in the establishment of INTELSAT, they were aware of the potential of satellite telecommunications and were keen to place the European industry in a position to benefit from this new technological adventure. In the mid 1960's a European Conference on Satellite Communications (CETS) launched the idea of a telecommunications satellite for Europe. The European Conference of Posts and Telecommunications (CEPT),[5] representing the Western European postal and telecommunications administrations, was then invited to

[2] G.A. Res. 1721 (XVI), at 21, U.N. Doc. A/4987 (Dec. 20, 1961). Available: http://www.oosa.unvienna.org/oosa/SpaceLaw/gares/html/gares_16_1721.html.

[3] International Telecommunications Satellite Organisation, established 1964. In this document "INTELSAT" refers to the satellite organisation prior to its restructuring and "ITSO" to the international organisation after restructuring.

[4] International Maritime Satellite Organisation, established 1975–79. In this document "INMARSAT" refers to the satellite organisation prior to restructuring and "IMSO" to the international organisation after restructuring.

[5] The CEPT is a forum of European representatives of administrations with no legal personality, no headquarters and no permanent secretariat. Available: http://www.cept.org/.

carry out a profitability study of a system of this kind. The study concluded that satellite links within Europe would never be economical in comparison with ground networks (which at the time were already at an advanced stage of development). It was felt that the distribution of television programmes was more promising, but the European Broadcasting Union (EBU),[6] in its turn regarded the project as far too expensive, and rejected it. The necessary political impulse came from the European Space Conference (ESC) in 1966[7] where it was decided to encourage research into a satellite solution for telephony and television. The ESC set up a working party in 1969, to which the CEPT associated itself.

The acronym EUTELSAT was first used in a technical study of the CEPT in 1971. This study recommended an innovative technological option for the overall design of a satellite system, using the Ku band for frequencies and digital technology for the transport of the telephony traffic (Time Division Multiple Access or TDMA). The CEPT accepted the principle of direct participation in the construction of a test satellite and a subsequent operational satellite, while requiring its member Post, Telegraph, and Telephone (PTT) administrations to study the feasibility of installing the required earth stations.

In 1976, the CEPT started negotiations with the European Space Agency (ESA)[8] for the implementation of a European telecommunications satellite system. It was decided that the Member States of ESA would fund in full the costs related to the space segment: four satellites to be launched successfully, launching services, earth equipment, in orbit control and monitoring of the satellites for a ten year period, starting in 1980. It was also decided that the users, that is the PTTs, would only contribute, via the organisation to be created for managing the system (EUTELSAT), through annual payments of an amount agreed in advance and that EUTELSAT would be the owner of the satellites and would have full freedom to use and operate them.

The scene was thus set for the establishment of EUTELSAT, with, as a first step, INTERIM EUTELSAT, from 1977–1985. Two main instruments

---

6 The EBU is an association under Swiss law, grouping together European public broadcasters for cooperation and exchange of audiovisual content. Available: http://www.ebu.ch/en/about/index.php.

7 European Space Conference, a ministerial conference held in Paris on 13 December 1966.

8 European Space Agency, established 1975. Available: http://www.esa.int/SPECIALS/About_ESA/SEM7VFEVL2F_0.html.

were put in place: the "Provisional Agreement"[9] and the "ECS Agreement."[10] The Provisional Agreement was not a treaty and INTERIM EUTELSAT had neither legal capacity nor personality. The French administration, as headquarters administration, was duly mandated to sign agreements and loans and to act in general in the name and on behalf of INTERIM EUTELSAT.

After the launch of an experimental flight unit in September 1977, the first satellite was launched in June 1983 and started its operational lifetime in October 1983 as EUTELSAT IF1, for the distribution of TV programmes to cable heads. The launch of a second satellite, EUTELSAT IF2, took place in August 1984. This satellite was dedicated to Eurovision transmissions for the EBU, telephony traffic and satellite multi-services (that is business services).

To proceed from INTERIM EUTELSAT to the permanent organisation, the Convention[11] and Operating Agreement of EUTELSAT were opened for signature in Paris in July 1982, in English and French languages, both texts being equally authentic, in a single original, and was signed on that date by 26 States. The Depository of the Convention was the Government of France. At an intergovernmental conference in Paris in December 1983, the Member States, aware that all the signatures and acts of notification, acceptance and approval, which were required for the entry into force of the Convention, might not be forthcoming by the 18 month deadline, decided to replace the 18 month deadline with a 36 month deadline. On 1 September 1985, the Convention and Operating Agreement entered into force in accordance with Article XXII of the Convention.

## 3. *Structure and Policy of EUTELSAT Prior to Restructuring*

In addition to the afore-mentioned Convention and Operating Agreement, as for any intergovernmental organisation, there was a Headquarters

[9] *Agreement on the Constitution of a Provisional European Telecommunications Satellite Organisation, "INTERIM EUTELSAT,"* agreed in Paris on 13 May 1977 between Administrations or Recognised Private Operating Agencies and deposited with the French Administration.

[10] *European Communications Satellite Agreement,* a supplementary agreement to the Provisional Agreement relating to the space segment of the satellite telecommunications system for the Fixed Service, agreed in Paris on 10 March 1978.

[11] Convention establishing the European Telecommunications Satellite Organisation (EUTELSAT), done 15 July 1982, entered into force 1 Sept. 1985, UK Misc. No. 4, Cmnd. 9154 (hereinafter EUTELSAT). Available: http://www.eutelsatigo.int/en/docs/ConventionB_en.pdf.

Agreement signed with France in November 1985 and a Protocol on Privileges and Immunities which entered into force in August 1988.

### 3.1. *Structure of EUTELSAT*

Like INTELSAT and INMARSAT, EUTELSAT was a two-tier organisation composed of an Assembly of Parties and a Board of Signatories, plus a permanent body, the executive organ, headed by a Director General.

The Assembly of Parties was composed of all the Parties to the Convention. Ordinary meetings were held every two years but extraordinary meetings could be held at the request of one or more Parties, supported by at least one third of the Parties, or at the request of the Board of Signatories, as happened in particular in the last years prior to restructuring. Each Party had one vote in the Assembly. Decisions on matters of substance were taken by an affirmative vote cast by at least two-thirds of the Parties present, or represented and voting. Decisions on procedural matters were taken by an affirmative vote cast by a simple majority. The functions of the Assembly of Parties were stated in Article IX of the Convention. In addition to decisions regarding accession/withdrawal of Parties, amendments to the Convention, relations with States and approval of the Headquarters Agreement, the Assembly's functions included the review of policy and objectives, authorisation of the use of the EUTELSAT Space Segment, and recommendations to the Board of Signatories on, inter alia, respect of treaties, proposed amendments to the Operating Agreement and use of separate space segment.

The Board of Signatories was composed of one representative of each and every Signatory. Unlike other organisations, such as INTELSAT, in the case of EUTELSAT this was manageable since it was a European regional organisation with a limited number of Signatories. Each Signatory participated in voting proportionate to its investment share, up to a limit of 20 percent of the total voting participation. The Board of Signatories tried to take decisions unanimously, but where this was not possible, decisions on matters of substance were taken in principle either by an affirmative vote of representatives of at least four Signatories holding at least two-thirds of the total voting participation, or by an affirmative vote cast by at least the total number of Signatories present or represented minus three, regardless of the voting participation represented by the latter. Decisions on procedural matters were taken by an affirmative vote cast by a single majority present and voting. In accordance with Article XII of the

Convention, the Board of Signatories, which, in practice, met every two months, was competent to take all the operational, technical, financial and commercial decisions. Furthermore, the Board of Signatories was competent to take decisions regarding the appointment and removal from office of the Director General, the conditions of employment of the staff of the Executive Organ and approval of the appointment of senior officers. The Board of Signatories submitted reports to the Assembly of Parties on the activities of EUTELSAT, was entitled to designate arbitrators in case of any arbitration involving EUTELSAT and made recommendations to the Assembly of Parties about proposed amendments to the Convention and Operating Agreement. The Board also examined applications for accession and made appropriate recommendations about these to the Assembly, in accordance with Article XXIII of the Convention.

As stated in Article XIII of the Convention, the Executive Organ, which was the permanent body of EUTELSAT, was headed by a Director General appointed by the Board of Signatories, subject to confirmation by the Parties. The Director General was the chief executive and legal representative of EUTELSAT, acting under the direction of the Board of Signatories and directly responsible to it for the performance of all functions of the Executive Organ. In practice, the Director General made proposals to the Board of Signatories and, with the support of the Executive Organ, ran EUTELSAT on a day-to-day basis in compliance with decisions taken by the Board of Signatories. The Director General was empowered to appoint all the staff of the Executive Organ.

### 3.2. *Activities of EUTELSAT*

The main purpose of EUTELSAT (as stated in Article III of the Convention) was the design, establishment, operation and maintenance of the space segment of the European telecommunications satellite system. Its prime objective was the provision of the space segment required for international public telecommunications services[12] in Europe. The EUTELSAT space

[12] "Public telecommunications services" were defined in Article I (k) of the Convention as fixed or mobile telecommunications services provided by satellite and available to the public, such as telephony, telegraphy, telex, facsimile, data transmission, videotex, transmission of radio and television programmes between approved earth stations having access to the EUTELSAT Space Segment for further transmission to the public and multi-services transmissions (closed network transmissions). EUTELSAT, Article I(k). Available: http://www.eutelsatigo.int/en/docs/ConventionB_en.pdf.

segment could also be made available for domestic public telecommunications services in Europe or to other domestic or international public telecommunications services outside Europe. Subject to appropriate terms and conditions, the EUTELSAT space segment could also be utilized in Europe for specialised telecommunications services,[13] but not for military purposes. In cases where the EUTELSAT space segment was used for domestic or international public telecommunications outside Europe or for specialized telecommunications services, or where space segment was provided separately from the EUTELSAT space segment, the Board of Signatories ensured that those Signatories which were not directly concerned would not have to participate in the financing.

The Convention made provision in Article XVI for the utilization of space segment equipment separate from the EUTELSAT space segment. An assessment had to be made of the potential significance of "economic harm" of a project of this nature as well as of its technical compatibility with the use of the radio frequency spectrum and orbital space by an existing or planned EUTELSAT space segment. It should be noted that Article XVI replicated equivalent provisions in INTELSAT's and INMARSAT's treaties, with which EUTELSAT had to comply.

The organisation functioned like a co-operative with Signatories contributing financially to the system in proportion to their investment shares, which were calculated in accordance with their use of the space segment (to which Article V(c) of the Convention and various articles of the Operating Agreement refer). In accordance with Article III(d) of the Convention, which states "in the implementation of its activities EUTELSAT shall apply the principle of non-discrimination as between Signatories," the Signatories provided services which were in competition with each other, using space segment capacity leased to EUTELSAT at cost. However, as specified in Article V(e)(i) of the Convention, the rates of utilization charged for each type of utilization had to be the same for all public or private telecommunications entities in territories under the jurisdiction of Parties, which had applied for space segment capacity for this purpose.

---

13 "Specialized telecommunications services" were defined in Article I (l) of the Convention as telecommunications services provided by satellite, other than public telecommunications services as defined above, including, but not limited to, radio-navigation services, broadcasting satellite services, space research services, meteorological services and remote sensing of earth resources. EUTELSAT, Article I(I). Available: http://www.eutelsatigo.int/en/docs/ConventionB_en.pdf.

This was irrespective of their geographical location or the level of their investment share. Signatories could only submit to EUTELSAT applications for allotment of EUTELSAT space segment capacity for territories under the jurisdiction of Parties. At the same time, as stated in Article V(b) of the Convention, the organisation had to "operate on a sound economic and financial basis having regard to accepted commercial principles." The revenues earned by EUTELSAT had to be used in the order of priority stated in the Operating Agreement, that is in the first place to meet operating, maintenance and administrative costs, then to repay capital to Signatories in proportion to their respective investment shares and finally to pay any remaining balance to Signatories as compensation for use of capital, again in proportion to their respective investment shares.

The procurement policy for the supply of goods and services (to which Article XIV of the Convention and an article of the Operating Agreement refer) was conducted on an international competitive basis, contracts being awarded to bidders offering the best combination of quality, price and delivery time. It was understood that in case of comparable bids, contracts would be awarded with due consideration to the general industrial interests of the Parties that is to manufacturers or providers of goods or services under the jurisdiction of a Party. In practice, for the procurement of satellites, despite being open international invitations to tender, the procurement contracts were nearly always awarded to Western European manufacturers. For the procurement of launching services, even though Ariane, the European launcher, was selected on many occasions, it was important for the organisation to have a double, even multi-source, policy to secure the continuity of services provided via EUTELSAT satellites.

As far as intellectual property rights were concerned, the policy of the organisation was, as stated in the Operating Agreement, to acquire only the rights which were necessary to enable work to be performed by or for EUTELSAT. Ownership of intellectual property generated by a contractor in the performance of a EUTELSAT-funded contract was retained by the contractor.

As regards liability, the Operating Agreement stated that there was a general provision to protect EUTELSAT and its Signatories against any liability or actions for damages to any Party, Signatory or to EUTELSAT itself, arising from any interruption, delay or malfunction of telecommunication services provided under the Convention or the Operating Agreement.

There was an institutional mechanism for the settlement of disputes, as mentioned in Article XX and Annex B to the Convention and also in the Operating Agreement, which potentially could involve the President of the International Court of Justice or the Secretary General of the Permanent Court of Arbitration. Fortunately, this was never required. The Operating Agreement also contained a provision for settlement of disputes regarding Allotment Agreements for space segment capacity. In the case of procurement contracts for satellites and launching services, there was generally a provision for arbitration in Paris and under French law, in accordance with the rules of arbitration of the International Chamber of Commerce.

### 3.3. *Evolution of EUTELSAT*

A few years after EUTELSAT came into being, there was much technological progress, in particular the possibility to operate more powerful satellites, for a longer operational lifetime (from 7 years for the EUTELSAT 1 satellites to 10–12 years for the second generation) with a greater number of transponders per satellite and subsequently with steerable antennae providing additional operational and commercial flexibility to respond to demand. On the earth equipment side, in addition to more efficient earth stations, reception antennae became smaller and less costly. Dual feed antennae allowed the direct reception of twice as many TV programmes transmitted by satellites operated at two different orbital locations. New EUTELSAT services developed in parallel, mainly in broadcasting, radio, data and business services as well as internet, localization and messaging, all of which proved to be profitable.

In 1992, the Board of Signatories took an historic decision of lasting importance to the success of EUTELSAT, which was the decision to allow the maximum possible number of satellites for Direct-to-Home (DTH) distribution of TV programmes to be placed at a single orbital location (firstly at 13° East, the "HOT BIRD"™ orbital location). This was the best way to enable users to receive a very large number of programmes without having to re-point their antenna. Prior to this, the Signatories favoured the principle of non-specialization of EUTELSAT satellites for the benefit of priority services, which could all be provided through any of the satellites of the EUTELSAT fleet. This decision permitted EUTELSAT to avoid a competitive disadvantage *vis-à-vis* its strongest rival, Société Européenne de

Satellites (SES), a private company established in Luxembourg, which already had adopted the strategy of co-positioning satellites for DTH distribution of TV programmes.

## 4. *Factors Influencing the Need for Restructuring*

Over the years, EUTELSAT had to face increasing competition in the field of satellite telecommunications. Right from the inception of EUTELSAT, there was competition from the terrestrial networks and submarine cables operated by the telecommunications administrations/public operators which held dominant, if not monopolistic, positions in their markets. In addition, there was competition between EUTELSAT and the other intergovernmental satellite organisations, INTELSAT and INMARSAT, even though their core businesses were different, with EUTELSAT's revenues increasingly derived from the provision of capacity for TV broadcasting and subsequently DTH distribution of TV programmes, that is not through cable heads but with direct reception via a satellite dish. There was also direct competition between EUTELSAT and some of its Signatories, as several of them at the same time operated national systems. For instance, in France, there were the Telecom I and II satellites, operating in the frequency bands of the Fixed Satellite Service (FSS), and the TDF satellites, operating in the Broadcasting Satellite Service (BSS) frequency bands of the International Telecommunication Union (ITU). In Germany, there were the equivalent telecommunications satellites, DFS Kopernikus and the broadcasting satellites TV.SAT; in Spain, Hispasat; in Turkey, Turksat, and in Greece, HellaSat. The fact that these Signatories were both investors in EUTELSAT as well as operators of their own national satellite systems meant that they were not necessarily practising fair competition. They had full visibility of the strategy and deployment plans of EUTELSAT through the documentation provided to them as members of the Board of Signatories and could influence decisions taken at the Board's meetings. EUTELSAT'S strongest competitor, as mentioned previously, was SES, which, since 1986, had been developing its market share for DTH distribution of TV programmes through its ASTRA satellites, firstly in Western Europe and later enlarging its geographical scope to Eastern and Central Europe and beyond.

Concerns about pricing and distribution policy arose from the fact that the Signatories were the sole distributors of EUTELSAT space segment capacity in their respective territories. There was no possibility for any other entity established in a country under the jurisdiction of a Party to obtain capacity direct from EUTELSAT. EUTELSAT, acting as wholesaler, was providing capacity to its Signatories at cost for their own use or for that of their distributors/retailers. The organisation was not authorised to sign direct contracts, except with entities outside the territories of EUTELSAT Parties, where a contract could be signed with a so-called Duly Authorised Telecommunications Entity (DATE), as provided for in the Operating Agreement. The Signatories preferred to make contracts with their customers for a service (that is capacity, plus the use and management of the technical equipment and premises) in order to maximise their profit in terms of mark-up over the payment of utilization charges to EUTELSAT. This was often done without a high degree of transparency, which raised concerns and complaints from customers for reasons which were obviously beyond the control of the organisation.

### 4.1. *European Commission's Satellite Green Paper*

In 1990, when the European Community was aiming to complete the creation of the Single Market by 1992, the European Commission produced its Satellite Green Paper.[14] The Commission believed that the compartmentalisation of the Community's satellite communications market might be contrary to the principles of the Single Market. The intention of the Green Paper was to extend the application of the generally agreed principles of Community telecommunications policy to satellite communications, taking into account the specificities of this means of communication. The Green Paper proposed four major changes:

- full liberalisation of the earth segment, including both receive only and receive/transmit terminals, subject to appropriate type approval and licensing procedures, where justified, to implement necessary regulatory safeguards;
- free unrestricted access to space segment capacity, subject to licensing procedures in order to safeguard the exclusive or special rights and regulatory provisions set up by Community Member States in conformity

14 "Towards Europe-wide systems and services – Green Paper on a common approach in the field of satellite communications in the European Community," Commission of the European Communities (CEC), accessed February 26, 2013. Available: http://europa.eu/documentation/official-docs/green-papers/pdf/satellite_communications_gp_com_90_490.pdf.

with Community law and based on the consensus achieved in Community telecommunications policy. Access should be on an equitable, non-discriminatory and cost-oriented basis;
- full commercial freedom for space segment providers, including direct marketing of satellite capacity to service providers and users, subject to compliance with the licensing procedures mentioned above and in conformity with Community law, in particular competition rules; and
- harmonisation measures as far as required to facilitate the provision of Europe-wide services. This concerned, in particular, the mutual recognition of licensing and type approval procedures, frequency coordination and coordination with regard to third country providers.

The Green Paper also defined the lines of action for creating a favourable environment, making a differentiation between EUTELSAT, INTELSAT and INMARSAT. As regards EUTELSAT, the Commission spoke of "working towards a review of the EUTELSAT Convention and its Operating Agreement as far as required to implement the proposed positions set out above",[15] whereas for INTELSAT and INMARSAT it spoke of "defining a common position in the international fora related to satellite communications in particular with regard to INTELSAT and INMARSAT".[16] For EUTELSAT, it recommended the following measures:

- "necessary modifications to ensure direct access for users to EUTELSAT satellite capacity;
- future handling of the co-ordination procedure concerning the "economic harm" provision of the EUTELSAT Convention in Article XVI(a);
- necessary measures for implementing future commercial independence of EUTELSAT;
- mechanisms for ensuring cost-orientation of the tariffs charged to users; and
- implementation of the separation of regulatory and operational interests and opening of membership of the EUTELSAT consortium to new parties."[17]

---

15 "*Towards Europe-wide systems and services – Green Paper on a common approach in the field of satellite communications in the European Community*," Section VI.4.1.

16 "*Towards Europe-wide systems and services – Green Paper on a common approach in the field of satellite communications in the European Community*," Section VI.4.2.

17 "*Towards Europe-wide systems and services – Green Paper on a common approach in the field of satellite communications in the European Community*," Section VI.4.1.

Community Member States held a total investment share of 88 percent in the EUTELSAT organisation, which meant, in the Commission's view, that they had to share responsibility for redirection and adjustment of the EUTELSAT Convention and Operating Agreement in order to ensure the development of the potential of EUTELSAT to the fullest possible extent, in line with the goals of the Single Market.

### 4.2. *EUTELSAT's Response to the Green Paper*

In December 1990, the Director General of DGXIII (Telecommunications) of the European Commission wrote to the Director General of EUTELSAT enclosing the official version of the Green Paper and inviting comments on the position and measures proposed therein, as described above. The EUTELSAT Assembly of Parties, Board of Signatories and Executive Organ all studied the document and their comments were forwarded to the Commission by the Assembly of Parties, following an extraordinary meeting of the Assembly in March 1991. A report from the Board of Signatories was attached as an Annex and it was pointed out that the Board of Signatories, as owners of the satellite system, played an essential role in the operation and functioning of the organisation.

In its reply to the European Commission, the Assembly of Parties contended that EUTELSAT had demonstrated its support for the objectives of the European Community through the promotion of European technology and the use of the European Currency Unit. They noted with satisfaction the recognition by the Commission of the organisation's special importance for Europe-wide satellite communications and for the continued development and growth of the pan-European infrastructure it provided. The Parties stated that any meaningful consideration of the proposals in the Green Paper had to take account of the unique position that EUTELSAT held in the European satellite telecommunications environment. This uniqueness derived primarily from its legal status. It was also pointed out that the organisation was subject to obligations, such as the provision of services on a non-discriminatory basis to all Signatories regardless of geographical situation and investment share, the policy of provision of services at minimum cost without regard to maximisation of profit and non-discrimination in the application of utilization charges. Since these obligations may have put EUTELSAT at a disadvantage *vis-à-vis* other entities which were providing satellite services in Europe, the question was raised whether it was logical, in respect of those services within the public telecommunications services mission, to equate EUTELSAT with those other entities that did not have the same status. Cognizant that only 12 of EUTELSAT's current 27 Member States were also Member States of the European Community, the Assembly of Parties, acknowledging the considerable flexibility inherent in the EUTELSAT statutes, considered that, under present circumstances, there appeared to be no need to amend the statutes. The Assembly of Parties recognised that some changes in practices and procedures might be needed, but it believed that, under the current circumstances, these could be achieved within the flexible framework of the Convention and Operating Agreement.

The Annex, containing the report of the Board of Signatories, addressed in detail and commented on the various proposals of the Green Paper, referring to the specific obligations of EUTELSAT and the existing separation of operational and regulatory functions. It pointed out the flexibility existing in the Convention and Operating Agreement, which meant that the provisions could in most cases be applied in a way that was consistent with the European Commission's views. The Annex also mentioned initiatives which had already been started in the existing framework of the organisation, for instance the introduction in 1990 of an expedited procedure for the application of Article XVI(a) of the Convention, which related to the question of "economic harm." The Annex also referred to the moves made by certain Member States regarding greater freedom of access to the EUTELSAT space segment.

## 5. *The Restructuring Process*

### 5.1. *Changes Made Under Existing Provisions*

Following the publication of the Green Paper, the organisation implemented a number of changes which were possible under its existing provisions. Following on from the above-mentioned expedited procedure for the application of Article XVI(a) of the Convention, the Assembly of Parties decided, between 1992 and 1994, to implement a simplified interpretation of this article, whereby a Signatory merely had to declare its intention to use a non-EUTELSAT space segment, as opposed to engaging in a consultation procedure. This ultimately rendered the consultation procedure regarding "economic harm" inapplicable.

The existing structure of the organisation was also able to accommodate changes relating to the provision of wider access to the EUTELSAT space segment. For instance, in 1995, to satisfy the growing number of service providers resulting from liberalisation in certain countries, the Assembly of Parties and the Board of Signatories authorised arrangements at the national level allowing Non-Signatory Entities (NSEs) to have direct access to the EUTELSAT space segment. NSEs were operators authorised by their national authorities, in consultation with the national Signatory, to have the same rights as the Signatory to access the EUTELSAT space segment, in proportion with utilisation. They, however, had no right of vote on the Board of Signatories. Another change which was accommodated within the existing structure was the implementation of multiple access arrangements, whereby a service provider could lease capacity from any Signatory in any of the countries party to this arrangement. These arrangements were open to all EUTELSAT Parties and Signatories and by 1999 ten countries had signed up. A further example was the establishment in the United Kingdom, in the late 1980's, of a Signatory Affairs Office, where requests for capacity from UK users could be managed through a structure within the UK Signatory, but separate from the commercial services of the official Signatory, British Telecom. A management fee of seven percent was payable to the Signatory over and above the cost of the space segment capacity. Finally, operators established in countries in which there was no Signatory (that is, countries which were not Party to the EUTELSAT Convention) could also have direct access to the EUTELSAT space segment as DATEs, subject to the agreement of the government of the country from which they uplinked to the EUTELSAT space segment. The first entity was authorised in 1992 and by 2000, there were over 30 DATEs.

Similarly, existing provisions of the Operating Agreement could be applied in a different way without the need to change the agreement itself. For instance, in order to facilitate the approval of earth stations, it was decided in 1993 to cease to apply the provisions of the article which stated that applications for approval of earth stations accessing the EUTELSAT space segment could only be submitted and approved by the Signatory in whose territory the earth station was located.

Following the first efforts, described above, for a smooth evolution of the organisation with new facilitating procedures, the Assembly of Parties accepted in 1995 the concept of having more than one Signatory per member country, which would entail amendments to Articles II, X, XI and XVIII of the Convention and made a formal decision to this effect in January 1996. It was decided at the same time to formally delete Article XVI, which as stated above was no longer applicable, and to delete Article XIX(c), so as to avoid a situation where an amendment was deemed null and void if it had not received formal acceptance of at least two-thirds of Member States whose Signatories held at least two-thirds of the total investment shares eighteen months after the date of its adoption by the Assembly, as stated in Article XIX(b). The underlying idea was to avoid potential unnecessary and indefinite repetition of amendment decisions by the Assembly, in particular if restructuring was at stake. In the event, none of these amendments to the Convention entered into force since an insufficient number of Member States were able to ratify them within the required eighteen months, despite repeated amendment procedures, before the July 2001 restructuring took place.

### 5.2. *Review of Potential Restructuring*

Meanwhile, the urgency for change was accelerating. The slow decision making process of the organisation, the visibility of its strategic decisions by some of its competitors, the fact that investment was linked to usage and not to development, together with the constraints on pricing and distribution, all had an impact on EUTELSAT's ability to face competition.

After looking at existing examples, for instance the Bank of International Settlements[18] and EUROFIMA,[19] it was decided that alternative models would need to be considered in EUTELSAT's case and a total of seven different models were studied in depth by two working groups established by the Parties and Signatories. These models included a scenario consisting only of a private company. After considerable discussion, the three following models were ultimately examined in great detail:

---

[18] Bank of International Settlements (BIS) was established in Basel (Switzerland) as a company with a legal status of an intergovernmental organisation with privileges and immunities and a headquarters' agreement.

[19] EUROFIMA, a European Company, was established in Basel by the Convention for the Financing of Railroad Rolling Stock, with by-laws attached to the Treaty.

- the continuation of EUTELSAT as a treaty-based organisation, with minor changes in its organic structure and more important changes in the distribution of functions and in the provisions relating to investment;
- a new intergovernmental treaty covering the continued involvement of Parties, together with the establishment of a national law company to which all assets would be transferred; and
- a continuation of the existing treaty together with the establishment of a national law company, with both entities making use of the space segment, but for different purposes.

Having agreed in 1996 upon the four "Basic Principles" of public/universal service, non-discrimination, fair competition and pan-European coverage of the satellite system, the Assembly of Parties agreed in May 1997 to the proposal of a new intergovernmental treaty enshrining these principles together with the establishment of a national law company, while noting that further study was needed on the maintenance of the interests of small Signatories, the European identity of the future company and the need for a "reasonable time for adjustment."

Meanwhile, in 1997, the organisation had commissioned a study by an international law expert[20] to examine the legal consequences of EUTELSAT restructuring itself by use of an amendment procedure, to establish whether this would be a practicable solution. The study concluded that EUTELSAT could only be restructured either by the conclusion of a new treaty or by a unanimous decision/consensus of the Member States to amend the existing treaty. This was based on the study's findings as follows:

- Amendment procedures for multilateral treaties require, unless anything else is stipulated in the treaty, the consent of all Member States.
- An amendment procedure laid down in a convention establishing an international organisation which requires a qualified majority and not unanimity must be understood on the basis of the limited jurisdiction attributed to the international organisation.
- The constitutions of many liberal democracies would not permit a state to join an international organisation with unlimited amendment procedures brought about by a qualified majority.

---

[20] This study is not publicly available.

- It is of no value to argue that states remaining in a minority could leave the international organisation because each member state has a full right to remain in the organisation and to be treated in line with the specific rules establishing the organisation.
- Member States of an international organisation, by a decision taken unanimously, can always change the charter of the international organisation, even without respecting the specific amendment procedures. This is a general rule of international treaty law. In such a case, since no minority exists, the question of violation of minority rights cannot arise.
- The system of amendment procedures and practice show that some treaties make a distinction between fundamental changes and other changes. No case has been found where a fundamental change completely restructuring the international organisation and changing its object and purpose has been brought about by a qualified majority against a minority.
- Although the doctrine does not frequently address the issue, important voices in the literature show that amendment procedures must be interpreted as being limited to amendments which remain within the general spectrum of object and purpose as well as the general structure of the international organisation.
- The structural changes and the removal of any operational responsibility from EUTELSAT by the envisaged amendment are fundamental and radical changes to the present international organisation EUTELSAT as well as to the Operating Agreement. A careful analysis shows that the EUTELSAT Convention does not permit a qualified majority, according to the amendment procedure, to bring about such a change. The EUTELSAT Member States could, however, implement such a decision by a unanimous vote and unanimous ratification.
- The termination of the EUTELSAT Operating Agreement cannot be brought about by a qualified majority decision against a minority. This view is reinforced by the special provision according to which the Operating Agreement must stay in force as long as the EUTELSAT Convention is in force.
- Although minority member states would have the legal possibility to immediately withdraw from EUTELSAT as soon as the qualified majority would bring into operation the envisaged amendment, they could also claim a violation of their rights under the EUTELSAT Convention and the Operating Agreement.

– Therefore, the envisaged changes to EUTELSAT can only be brought about by the conclusion of a new treaty or by a unanimous decision/consensus of the member states.

### 5.3. *Fixing a Date for Restructuring*

The question of securing a fixed date for the entry into operation of EUTELSAT's restructuring was given lengthy consideration. The solution proposed was the one already adopted for INMARSAT, that is provisional application of amendments to the Convention, as provided for in Article 25 of the Vienna Convention on the Law of Treaties.[21] In the case of EUTELSAT, in accordance with the cumbersome procedure contained in Article XIX of its Convention, no amendment could enter into force until the Depositary had received acceptance of the amendment(s) from two-thirds of the Parties whose Signatories represented at least two-thirds of the investment shares. It was considered essential to have a specific date for the entry into force of EUTELSAT restructuring and thus, in parallel to the "normal" amendment procedure, which allowed a reasonable time for Parties to notify their acceptance of the amendments, a deadline for restructuring would be set. If the required number of acceptances were not reached by the deadline, the amendments would nevertheless be applied provisionally. To avoid legal difficulties of a constitutional nature in certain countries, it was planned to insert some provisions in the decisions to be taken.

Provisional application of treaties has not been the subject of many specific studies even though this concept has been considered a number of times since 1945. In the list of treaties published by the UN Secretary General,[22] there are around one thousand treaties which foresee provisional application in a wide range of domains. In the case of EUTELSAT, it was not envisaged that Member States would decide unilaterally to enforce a treaty in advance, but that mutually acceptable provisions of a contractual nature would be put in place. Provided there was reciprocity, these would then constitute acts creating rights and obligations for those States. Article 24 of the Vienna Convention was also relevant, since it provides,

---

21 Vienna Convention on the Law of Treaties, art. 25, *done* May 23, 1969, 1155 U.N.T.S. 331. Available: http://untreaty.un.org/ilc/texts/instruments/english/conventions/1_1_1969.pdf.

22 "List of multilateral treaties deposited with the Secretary-General," United Nations, last accessed Feb. 27, 2013. Available: http://www.un.org/millennium/law/titles.htm.

with some flexibility, for Member States to decide on the entry into force of the treaty they have adopted. From this, as regards provisional application of amendments to the EUTELSAT Convention, it could be concluded as follows:

- States might decide to provisionally enforce a treaty or some of its provisions: there are no formalities to such acceptance, evidence of their will is sufficient;
- they could put an end to provisional application by indicating that they are not in favour of proceeding to ratification; but
- they would nevertheless be committed to abide to obligations they have accepted.

It was of paramount importance that the need for urgency would be recognised by all Member States by consensus. From a legal point of view, it was checked that the principles as stated in Article 25 of the Vienna Convention were applicable to treaties creating intergovernmental organisations, including amendments to such a treaty, and that a decision for provisional application might be taken through a resolution by an intergovernmental organisation. As a result, all Member States would be obliged to implement a provisional application of amendments to the Convention, provided that the decision to do so had been adopted by consensus. Any Member States which had not ratified the amendments by the prescribed deadline would have to enforce, in good faith, the amending decisions to the extent that national rules of constitutional, legislative or regulatory nature would not prevent them from so doing. The Member States which had ratified the amendments by the prescribed deadline could not question their obligation to enforce the amending decisions unless they were to use their right to withdraw from the organisation.

### 5.4. *Maintenance of a European Identity*

There were several legal issues related to the wishes of a number of Parties and Signatories to retain a European identity for the company which would be established under national law, notably in terms of contracts for goods or services and recruitment and working language(s). EUTELSAT had a certain number of European characteristics, since it was foreseen as a tool for European co-operation and integration. Only European States could be Member States (with a wide interpretation of Europe encompassing not only Western Europe but also Central and Eastern Europe, as well as countries of the former USSR). The Signatories (as partners, distributors and investors) were entities established in the territory of a Member State, that is in "Europe," and the core business of EUTELSAT was situated in Europe. Furthermore, the pan-European satellite system was providing, *inter alia*, public services on a European scale, thus making a definite contribution to the establishment of trans-European networks and infrastructure, in the economic interests of Europe. Satellites were, in practice, ordered from European suppliers even though there was an international open tender process each time. The same applied for the vast majority of launching services, as well as for other products and services provided to the EUTELSAT organisation. It could however be seen as tantamount to unlawful discriminatory practice under, for instance, French law if it were stated in the Articles of Association (which is a public document) that the company "encourages the employment of citizens of European countries." This could even involve criminal liability for the managers of the company and the company itself. Furthermore it could be deemed incompatible with French law if the Articles of Association stated that the working languages of the company are French and English and that documents required for the General Meeting of Shareholders, the Supervisory Board and the Management Board should be in both languages (the Toubon Law, mandating the use of the French language in all official and public documentation in France). It was therefore recommended that this provision, instead of being stated in the Articles of Association, be reaffirmed by the company's managers and directors as general policy and included in internal procedures. The only reference to Europe in the Articles of Association which could be included related to procurement, by adapting the principles already in place in Article XIV of the Convention and in the Operating Agreement.

In the Amended EUTELSAT Convention,[23] inclusion of an explicit reference to Europe in the preamble was proposed, in particular in the fourth paragraph "wishing to continue the establishment and operation of the EUTELSAT telecommunications satellite system as part of a trans-European telecommunications network for providing telecommunications

23 Convention Establishing the European Telecommunications Satellite Organisation, done 15 July 1982, entered into force 1 Sept. 1985, UK Misc. No. 4, Cmnd. 9154. (amended 20 May 1999, amended version applied provisionally 2 July 2001, entered into force 28 November 2002). Available: http://www.eutelsatigo.int/en/docs/Amended_Convention.pdf.

services to all participating States, without prejudice to the rights and obligations of the States parties to the relevant European Union (EU) and other international agreements." In addition, the Basic Principles, (see Article III(a)(ii) of the Amended EUTELSAT Convention) would include "pan-European coverage by the satellite system: the company Eutelsat S.A. shall, on an economic basis, seek through the pan-European coverage of its satellite system to serve all areas where there is a need for communications services in Member States." This Basic Principle is indeed fundamental in EUTELSAT's restructuring. Finally, the provisions restricting membership of the EUTELSAT organisation to European States would remain.

### 5.5. *Protection of Small Signatories*

One of the key elements which attracted Member States to EUTELSAT was the fact that all Signatories would be treated equitably and would benefit from non-discriminatory treatment, regarding, in particular, access to all services and applications provided by EUTELSAT through its space segment, regardless of their geographical location in Europe or their level of investment share. The protection of small Signatories/minority shareholders was raised early on in the restructuring deliberations. The aim was to ensure the continuation of non-discriminatory treatment following restructuring and also to ensure a balanced representation of small Signatories/small investors in the bodies of the restructured EUTELSAT. It was ultimately decided to address this issue by ensuring that all groups of Signatories would be represented on the Board of the national law company. This was done by dividing the Signatories into three groups, based on their level of use of the EUTELSAT space segment, which was then translated into a proportionate level of investment shares.

### 5.6. *Approval of Restructuring*

After seven years of deliberations on the subject, a restructuring package for EUTELSAT was finally approved by consensus of the Assembly of Parties in May 1999. This consisted of an amended Convention, containing the Basic Principles, and the transfer of all assets, liabilities and operational activities of EUTELSAT to a private company to be established under French law. With regard to the Convention and Operating Agreement, the Assembly approved the amendments proposed by the Party of France as host Party, together with additions suggested by the Board of Signatories and the Joint Parties' and Signatories' Study Group. It also noted that the Operating Agreement would be terminated upon entry into force of the Amended EUTELSAT Convention. The Assembly also approved the text of the Arrangement between EUTELSAT IGO and Eutelsat S.A., noting that it had previously been approved by the Board of Signatories and that it would be signed once Eutelsat S.A. was established. Noting the draft text of the Articles of Association of Eutelsat S.A., the Parties agreed that this took into account the general principles they had sought.

At the same meeting, the Assembly approved in principle the text of the Transfer Agreement, noting that the Board of Signatories had already done so. It also noted that the Transfer Agreement would be finalised later and signed by the Executive Secretary of EUTELSAT IGO and the Chairman of the Directorate of Eutelsat S.A. The Assembly declared that the transfer would be implemented by the method described in Annex A of the Amended EUTELSAT Convention and in accordance with the provisions of the Transfer Agreement, that is that EUTELSAT would enter into a Transfer Agreement with the company Eutelsat S.A. so as to transfer to the company all its assets and all liabilities relating to its activity, that this would result in a universal transfer of all rights and obligations, and the transferred assets and liabilities would constitute a complete and autonomous branch of activity. Such a transfer would be similar to a spinoff ("scission") transfer of activity, provided under Article 382 and following the French law relating to commercial companies,[24] except that none of the obligations normally binding upon the transferring party under this law would be applicable to EUTELSAT. It was agreed that the transfer would take effect *erga omnes*, as of 2 July 2001. In exchange for the transfer, EUTELSAT would receive shares in the company Eutelsat S.A. These shares would be passed on immediately and unconditionally to the Signatories as new shareholders of the company, in proportion to their respective existing holdings of investment shares in EUTELSAT, as determined at a date to be on, or immediately before, the date of the transfer.

The taxation conditions applicable to the transfer were noted by the Assembly of Parties, as well as those applicable to Eutelsat S.A. for a transitional period. It was noted that actions had already been initiated for the orderly transfer of assets and liabilities from EUTELSAT to the company

---

24 Act No. 66-537 of 24 July 1966 on commercial companies. Available in French only: http://www.legifrance.gouv.fr/affichTexte.do;jsessionid=B20B6EF20F9AE02C6B9B9E24597BBB31.tpdjo09v_2?cidTexte=JORFTEXT000000692245&dateTexte=20090924.

and that there would be no recourse to Parties, Signatories or the intergovernmental organisation once the company was established. Liability for outstanding loans and notes of EUTELSAT would be assumed by Eutelsat S.A. in accordance with the Transfer Agreement.

The principles as enunciated in Annex A to the Amended EUTELSAT Convention and in the Arrangement were the essential components of the restructuring package as far as staff and retired staff of the Executive Organ were concerned. Furthermore, the Assembly noted the report of the Board of Signatories on the elements of the Business Plan of Eutelsat S.A.

Since there was no amendment procedure specified in the Protocol on Privileges and Immunities, the Director General proposed that the Assembly should adopt the text of an amended Protocol to apply to the intergovernmental organisation, the representatives of Parties and the Executive Secretary and his/her staff, after restructuring had taken place. In accordance with Article XII(c) of the draft text of the amended Convention, the Parties approved the text of an amended Protocol between EUTELSAT IGO and the EUTELSAT Member States and noted that an Amending Agreement to the Protocol would be signed on behalf of EUTELSAT IGO by the Executive Secretary or by the Director General acting as Executive Secretary.

The Assembly of Parties also noted that the Director General and the French Government were negotiating an amended Headquarters Agreement to apply to the intergovernmental organisation, the representatives of Parties and the Executive Secretary and his/her staff, after restructuring had taken place. In accordance with Article XII(c) of the Amended EUTELSAT Convention, the draft amended Headquarters Agreement between EUTELSAT IGO and the French Government was approved, subject to any modifications of a technical or non-substantive nature which might be agreed between the Director General and the Government before the signing of the amended Headquarters Agreement. It was noted that this would be signed on behalf of EUTELSAT IGO either by the Executive Secretary or by the Director General acting as Executive Secretary and that it would take effect on transition to the new structure, replacing the Headquarters Agreement of November 1985.

The Assembly adopted a Resolution in May 1999 on accelerated (provisional) implementation, in which it noted the following conclusions of the Board of Signatories:[25]

---

[25] This Resolution is not publicly available.

- the vital importance for EUTELSAT to achieve restructuring within a fixed time-frame;
- unless restructuring were implemented in an early time-frame, the future of EUTELSAT could be seriously affected, given the increasingly aggressive competition; moreover, once a decision to restructure had been taken by consensus, it would be neither sound nor logical to prolong a transitional situation indefinitely, not only in terms of EUTELSAT's image but also with regard to the market and the financial community, that is the maintenance of confidence being paramount if EUTELSAT were to raise additional capital at the lowest possible cost and to maintain and build its customer base;
- the introduction of a programme of actions by the Director General, once the amendments on restructuring had been approved by the Assembly, so as to encourage Parties to ratify the amendments and give them all the assistance they required in completing their domestic procedures, so that the necessary number of ratifications could be obtained for entry into force by 29 September 2000 (this deadline was finally postponed to 2 July 2001); and
- the need for EUTELSAT to have available an alternative solution for accelerated implementation of the restructuring within a reasonable time-frame, should the number of ratifications for entry into force of the amendments not be obtained before the date mentioned above.

### 5.7. *Agreement of the European Commission*

To permit the restructuring of the organisation, it was necessary to inform the European Commission, in accordance with Article 2 and 4 of (at the time EEC) Council Regulation 17/62,[26] and obtain its approval. This procedure is called "negative clearance." EUTELSAT submitted an application for negative clearance and/or notification for an individual exemption in July 1999. The submission contained information on the principal business of the organisation, its overall turnover, the system of contributions proportionate to investment shares of the Signatories (including details of

[26] Council Regulation No 17, First Regulation implementing Articles 85 and 86 of the Treaty, 1962 O.J. (13). Available: http://eur-lex.europa.eu/LexUriServ/LexUriServ.do?uri=CELEX:31962R0017:EN:HTML.

investment share distribution for each Signatory), the urgency of the need to transform EUTELSAT into a structure which would allow it to meet the increasingly fierce competition, develop its service offerings, enter into partnership and obtain flexibility to expand the its financial base. Emphasis was placed on the decisions taken by the Assembly of Parties in May 1999 including, if necessary, an accelerated procedure for implementation of the restructuring project ensuring an implementation date of 2 July 2001 at the latest. In the document, it was specified that, if, for reasons of time, it would not be possible for the Commission to decide formally on this case, EUTELSAT would not oppose the notification being dealt by an administrative letter. EUTELSAT attached the drafts of the Amended EUTELSAT Convention, the Arrangement, the Articles of Association of the company, the Agreement amending the Protocol on Privileges and Immunities and the general framework guiding the distribution policy to be adopted by the company. The application stated that the restructuring of EUTELSAT was aimed at enabling it to respond to the requirements of the newly legalised European Community telecommunications market, meeting thereby the concerns expressed by the Commission and the Council on this issue. As regards the question of possible restrictions on the freedom of the company to take independent commercial decisions, the Commission was advised that nothing in the notified proposals would significantly restrict this freedom. Although it should be noted that it was specified in the Amended EUTELSAT Convention that EUTELSAT IGO would ensure that Eutelsat S.A. respect certain Basic Principles, nevertheless the organisation would in no way interfere with the commercial activities of Eutelsat S.A.

A further point covered the possible effect on trade of Member States of the Community and/or European Free Trade Association (EFTA) States. It was pointed out that the primary objective of EUTELSAT was to provide the satellite capacity necessary for international telecommunications services in Europe and that all Member States of the European Community and EFTA were Parties to the Convention and had designated a Signatory which, following restructuring, would become a shareholder in Eutelsat S.A. After restructuring, Eutelsat S.A. would carry out the operational activities of EUTELSAT. Consequently, given the geographic zone covered by EUTELSAT and the cross-border nature of telecommunications services offered via its satellites, the restructuring of EUTELSAT would be unlikely to affect trade both between Member States of the European Community and Member States of EFTA and other States Parties to the EUTELSAT Convention. In particular the restructuring of EUTELSAT would be likely to lead to increased competition at both an intra-brand and inter-brand level on the relevant product markets affected by it, thus leading to new patterns of trade within the territory of the European Economic Area.

EUTELSAT provided the European Commission with information about the relevant product market, defined in this case as being the market for satellite transmission, the relevant geographic market, the position of participants in the affected relevant product markets, the market for the provision of satellite transmission capacity, and the various activities for which EUTELSAT capacity was used, such as DTH services. Information was also provided on the other activities for which EUTELSAT capacity was used, that is corporate telecommunications networks and very small aperture terminal (VSAT) networks and for mobile messaging on personal communications services by satellite. In addition, elements were given as to the position of competitors (mainly SES ASTRA, INTELSAT, TELENOR, HISPASAT, ORION, PANAMSAT and SIRIUS) and main customers (British Telecommunications, Telecom Italia, France Télécom, Deutsche Telekom and the Royal Post and Telecommunications of the Netherlands (KPN)) on the relevant market, market entry and potential competition in product/geographic areas. Finally information was provided on the reasons for the application for negative clearance as well as the reasons for the application for exemption under Article 85(3) of the European Community Treaty[27] at the time.

The answer to the submission came from the European Commission in the form of a negative clearance letter in November 2000. This stated that EUTELSAT restructuring was not in the Commission's view affected by Article 81(1) of the European Community Treaty. The European Commission noted that an Initial Public Offering (IPO) of 30 percent of the shares of Eutelsat S.A. was planned within two years of the restructuring (that is by July 2003) and that, in the Commission's view, this would dilute the shareholdings of the former Signatories and would also reduce the possible conflict of interest between their role as shareholders and as distributors of Eutelsat S.A.'s services. It would also unambiguously permit the direct

---

[27] Treaty Establishing the European Economic Community, 25 March 1957, 298 U.N.T.S. 3. Available: http://eur-lex.europa.eu/en/treaties/dat/12002E/htm/C_2002325EN.003301.html.

distribution of services by Eutelsat S.A. rather than requiring customers to approach Eutelsat S.A.'s current distributors (primarily the Signatories).[28]

5.8. *Process of Restructuring*

For the restructuring of EUTELSAT, it was decided to set up a société anonyme incorporated under French law (Eutelsat S.A.) with a Supervisory Board and a Management Board. The business would be transferred via a spin-off to Eutelsat S.A. of all the organisation's assets and liabilities (transfert universel du patrimoine d'EUTELSAT) together with its rights and obligations. Legally speaking, all transfer operations would be completed as of 2 July 2001. All these operations and implementing procedures were reviewed by the Eutelsat S.A. Supervisory Board at meetings in May 2001.

The EUTELSAT organisation bought a shell company with a registered capital of 40,000 euros divided into 2,500 shares of 16 euros each. The organisation became the sole shareholder in the shell company, Eutelsat S.A., in March 2001. This company conducted no operational activity prior to 2 July 2001. By decision of its sole shareholder at an Extraordinary General Meeting, also in March 2001, Eutelsat S.A. was transformed into a société anonyme with a Supervisory Board and a Management Board. The Director of Legal Affairs of the organisation was appointed member of the Management Board and Sole Director, with the specific task of completing the legal formalities prior to the transfer of EUTELSAT's business, on behalf of the Eutelsat S.A. shell company. In preparation for the events of 2 July 2001, the sole shareholder decided to modify the share capital of the company by dividing the existing shares into 40,000 shares with a value of 1 euro each by a decision taken in June 2001. To meet the obligations imposed by French law, the EUTELSAT organisation lent one share in the company to each member of the Supervisory Board, such loan terminating on 2 July 2001. Thus, prior to 2 July 2001, the EUTELSAT organisation held 39,985 shares in the company.

The main legal events which took place at an Extraordinary and an Ordinary General Meeting of the company's shareholders on the morning

---

[28] "Commission gives green light to EUTELSAT restructuring," European Union, accessed Feb. 27, 2013. Available: http://europa.eu/rapid/pressReleasesAction.do?reference=IP/00/1360&format=HTML&aged=0&language=EN&guiLanguage=en.

of 2 July, convened on 15 June 2001 in accordance with French law, were as follows (in chronological order):

- the Supervisory Board appointed the members and Chairman of the company's Management Board, as the mandate of the Sole Director had ended;
- the shareholders, and particularly the principal shareholder, EUTELSAT, approved the Transfer Agreement;
- the shareholders, and particularly the principal shareholder, approved the corresponding changes to the present Articles of Association of Eutelsat S.A.;
- the shareholders, and particularly the principal shareholder, acknowledged the resignation of the Sole Director, whose mandate had come to an end, and the appointment of the new members and Chairman of the company's Management Board;
- Eutelsat S.A. and the EUTELSAT organisation had to sign the Transfer Agreement transferring the organisation's business and its assets/liabilities;
- once the Transfer Agreement had been signed by Eutelsat S.A. and the organisation, the organisation transferred to each shareholder the number of shares in the company to which it was entitled;
- Eutelsat S.A. and EUTELSAT IGO signed the Arrangement.

## 6. *The Operating Company EUTELSAT S.A.*

In the ten years since it was established, Eutelsat S.A. has been highly successful, with successive annual growth, as evidenced by its financial results, and is the third largest satellite operator in the world. Its structure has evolved over the years to meet the demands of the changing economic environment.

### 6.1. *Structure of Eutelsat*

At the start, in 2001, Eutelsat S.A. was a joint stock company with limited liabilities under French law. Its governance was initially composed of a Supervisory Board of 15 members and a Directorate of four persons, headed by a Chairman who was also Chief Executive Officer of the company. As mentioned earlier, the composition of the Supervisory Board was designed to ensure the ongoing representation of small Signatories/minority shareholders in the governance of Eutelsat S.A. In July 2001, it consisted of ten Board members representing former Signatories which held over seven percent of investment shares (three from France Telecom, three from British Telecommunications (BT), three from Telecom Italia, and one from Deutsche Telekom), three Board members representing former Signatories which held between one and seven percent of share capital and two Board members representing former Signatories with under one percent of shares. This legal structure of a joint stock company with a Supervisory Board and a Directorate had been previously agreed upon to accommodate the strong wish of the former Signatories to replicate the earlier structure which consisted of a Board of Signatories and an executive organ headed by a Director General. They concurred that a structure with a Board of Directors and a CEO/Chairman of the Board would concentrate power in the hands of the management, and more precisely into those of the CEO/ Chairman, and wished to avoid this. The Signatories did not realise that under French law, a private company would have autonomy *vis-à-vis* its shareholders. They believed that the structure put in place, as supported by the German Signatory, would closely replicate that of an "Aktiengesellshaft" (joint stock company) with an "Aufsichtsrat" (Supervisory Board) and a "Vorstand" (Directorate). German law gives much more power of intervention for the Supervisory Board than is the case under French law. Under French law, the Directorate manages the company under the supervision of the Supervisory Board, which in practice means that the role of the Supervisory Board is mainly to monitor the activities already undertaken by the Directorate, except for some decisions where prior agreement from the Board is needed.

At the General Assembly of Shareholders in November 2004, it was decided to change the Articles of Association of Eutelsat S.A. to provide a structure with a Board of Directors and a CEO who was at the same time Chairman of the Board of Directors. This was considered more appropriate and more efficient in terms of the balance of powers within the governing bodies of the company.

In 2005, Eutelsat Communications S.A. became the holding company of Eutelsat S.A. and the holding company was put into an IPO. Both companies continued with the system of governance described in the previous paragraph until November 2009, when the General Assemblies of Shareholders of Eutelsat S.A. and Eutelsat Communications S.A. decided to separate the functions of CEO and Chairman of the Board of Directors and to allow the presence and participation of independent members in meetings of the Board of Directors. The former CEO was replaced, although he has, up to now, continued to serve as the Chairman of the Board of Directors of both Eutelsat S.A. and Eutelsat Communications S.A.

The Articles of Association of Eutelsat S.A. originally contained several specific provisions which differentiated it from a standard company. One of these provisions, which is ongoing, is the reference in the Articles of Association to the Amended EUTELSAT Convention, which means that the activities of the company must comply with the four Basic Principles embodied in the Amended Convention. Another ongoing provision states that in observance of EU rules and relevant international agreements, contracts for goods and services will, as far as possible, be awarded with consideration to the interests of European industry if there are bids of comparable quality, price, delivery time and other important criteria of relevance to the company. Prior to the establishment of the holding company in 2005, there were limitations on voting rights, with shareholders only able to have a maximum of 20 percent of voting rights regardless of the size of their shareholding. This replicated a similar limitation which existed in the organisation prior to restructuring. Until 2004, the Articles of Association also specified the composition of the Supervisory Board, as described above.

The Board of Directors of Eutelsat Communications is further assisted in its task by a Selection and Remuneration Committee, an Audit Committee and a Strategy and Investment Committee.

Eutelsat Communications, the holding company, retains a 96.11 percent stake in Eutelsat S.A. Since 2001, Eutelsat has created subsidiaries which, as their principal activities, promote products and services and represent Eutelsat S.A. in the United States, Brazil, Great Britain, Poland, Italy, Malta, Turkey, Spain, Portugal, Germany and France. In addition, through its subsidiary in Germany, the Eutelsat Group holds 27.69 percent of the Spanish satellite operator Hispasat. Eutelsat also holds 50 percent of Solaris Mobile Ltd, a company incorporated in Ireland and jointly owned with SES Astra S.A.[29]

---

29 Solaris Mobile has the specific role of operating and marketing the S-band payload of Eutelsat W2A satellite in compliance with the authorisation granted to it by the European Commission in May 2009.

### 6.2. *Shareholding*

Eutelsat S.A. started operations on 2 July 2001, soon after the collapse of the internet bubble. Under these complicated economic conditions, the large telecommunications operators, which had become shareholders of the company, found themselves in a fragile position and, despite the importance of their initial investment in Eutelsat, most of them were forced to sell their shares. The decision to sell came as a surprise and raised the question as to who would be in a position to buy their shares: industrial entities, media companies, financial institutions or banks. It should be remembered that the European Commission, when it had studied EUTELSAT's proposals for restructuring, had been very sensitive to the potential risk of anti-competitive practices and/or abuse of a dominant position from the big telecommunications operators, namely. France Telecom, British Telecommunications, Telecom Italia and Deutsche Telekom, and it requested an IPO for Eutelsat S.A. of at least 30 percent of its shares, so as to open up the shareholding structure to new entities. When the telecommunications operators started to sell, the only entities able or willing to invest in Eutelsat S.A. were financial investors. Between 2002 and 2005 all the major telecommunication operators sold their shares to financial investors which then resold them to other financial investors. In 2005, as mentioned below, following a leverage recapitalisation and the creation of the holding company Eutelsat Communications S.A., more than 95 percent of the shareholding of Eutelsat S.A. was held by the holding company, which was floated on the Paris Stock exchange. In January and February 2007, two long term investors, namely the Spanish group Abertis and the French entity CDC ("Caisse des dépôts et consignations") became major shareholders. To date, the ownership structure of Eutelsat S.A. has remained unchanged. As regards the current situation for Eutelsat Communications S.A., Abertis Telecom holds 31.36 percent, "Fonds Stratégique d'investissement" (FSI) holds 25.62 percent (CDC transferred its shareholding to its subsidiary FSI, a fund co-owned by the French State in July 2009) and there is a free float of 38.05 percent, with remaining shares allocated to Directors and employees.

### 6.3. *Leverage Recapitalisation and Initial Public Offering*

In 2004, the majority shareholders of Eutelsat S.A. set in motion a chain of events to amend the Articles of Association in order to implement a leverage recapitalisation of the company, with the aim of setting up a holding company which would become the controlling shareholder of Eutelsat S.A. The holding company they established was incorporated under French law in February 2005 and, in August 2005, became Eutelsat Communications S.A. In April 2005, the holding company acquired Eutelsat S.A. In order to allow the minority shareholders of Eutelsat S.A. to participate in the recapitalisation, a proposal was made to them to either exchange their interest in Eutelsat S.A. into an interest in the holding company, under the same conditions and, in particular, with the same value as those offered to the majority shareholders, or to sell their shares in Eutelsat S.A. The majority shareholders who had initiated the leverage recapitalisation notified EUTELSAT IGO that they would like to proceed with the listing of Eutelsat Communications S.A. on the Paris stock exchange instead of Eutelsat S.A. and explained their rationale for doing this. They wished to ensure that the flotation would be as large as possible and, as a result, that the percentage of Eutelsat Communications S.A.'s share capital held by the public would be greater than that which would have been obtained if Eutelsat S.A. were to be listed. They also pointed out that new investors buying into the share capital of Eutelsat Communications S.A. after an introduction on the stock exchange would be able to profit from the effect of the financial leverage obtained by the group of companies (that is, Eutelsat Communications S.A. and its subsidiaries). They added that the listing of Eutelsat Communications S.A. was the only way to offer direct liquidity to the investors in Eutelsat S.A. who were shareholders of Eutelsat Communications S.A. prior to its public flotation.

Listing on the stock exchange had previously been considered on several occasions. In 2003, the listing of Eutelsat S.A. was aborted due to adverse market conditions and attempts were made on two further occasions in the autumn of 2005. Eutelsat Communications was finally listed on the Paris stock exchange on 2 December 2005 with a value per share of 12 euros. (The current value per share is around 30 euros).

### 6.4. *Share Value*

Prior to restructuring, EUTELSAT benefited from an excellent rating at a level normally reserved to States. This was mainly due to the structure of the organisation itself. Firstly, the investors were telecommunications operators, which at the time were in a dominant position in their respective territories, being either companies or parts of Ministries and therefore strong and healthy. Secondly, the investors were designated by Member States. The presence and role of Parties gave an additional layer of comfort to the banks and financial institutions when there was a need for EUTELSAT to go to the financial markets in order to obtain loans. After restructuring, the rating changed drastically, becoming even more dramatic in the wake of the collapse of the internet bubble, which still had a strong impact on the information technology sector. The telecommunications operators which were the main shareholders of Eutelsat S.A. had very large debts and were obliged to disinvest from Eutelsat S.A. within a two year period. As of today, after this period of turmoil and the ensuing major changes in the shareholding structure, Eutelsat is rated by Moody's and Standard and Poors as "investment grade" on the basis of the robustness of its business and its excellent results year after year. This rating allows Eutelsat Communications and Eutelsat S.A. to obtain very good conditions for loans or refinancing from the financial entities when required.

### 6.5. *Regulatory Control*

As an operator providing satellite capacity and services to a number of countries, Eutelsat S.A. is subject to strict regulatory controls. Not only must it comply with the national regulations of its home country, France, and all other countries in which it provides capacity and services, but its operations are also governed by international regulations with which these countries themselves must comply.

Amongst these are the national regulations governing access to radio frequency spectrum and various related authorisations ("frequency assignments") and international regulations governing the co-ordination of these authorisations at the international level, as defined in the Radio Regulations of the ITU.[30] Most of the frequency assignments used by Eutelsat S.A. in its current activities, and those of the foreseeable future, were issued to the EUTELSAT Member States collectively prior to restructuring. EUTELSAT IGO continues to hold the joint rights and obligations relating to these collective frequency assignments. Eutelsat S.A. has simply succeeded EUTELSAT as the satellite operator authorised to conduct operations under these frequency assignments and does so pursuant to the rights

30 "Radio Regulations," International Telecommunication Union, accessed Feb. 27, 2013. Available: http://www.itu.int/pub/R-REG-RR/en.

contained in the Arrangement between EUTELSAT IGO and Eutelsat S.A. Frequency assignments which have been issued since the restructuring are assigned by the ITU to the French authority who then assigns them directly to Eutelsat S.A. In both cases, that is where the frequency rights are held by EUTELSAT IGO or by Eutelsat S.A., they are co-ordinated by the Agence nationale des fréquences (ANFR), the French national authority responsible for ensuring that France complies with its obligations under the ITU Radio Regulations. Satellite frequency assignments held directly by Eutelsat S.A. have, since 2004, been subject to a law[31] which requires any operations under these assignments to be further authorised by the Minister for Electronic Communications, following advice from various authorities (Conseil supérieur de l'audiovisuel, Regulatory Authority for Electronic Communication and Posts (ARCEP) Ministry of Defence). Similarly, in the provision of space segment capacity to entities in approximately 150 countries and territories, Eutelsat S.A. is subject to the national communications and broadcasting laws and regulations of many countries.

In July 2001, the French Minister of Telecommunications granted authorisation for Eutelsat S.A. to establish and operate a telecommunications satellite network open to the general public in France for services other than public telephony for a period of 15 years. Since July 2004,[32] the installation and operation of networks open to the general public are now unrestricted, provided a prior declaration is made to ARCEP, the authority responsible for ensuring that operators comply with the obligations contained in applicable legislative and regulatory requirements in France. These activities still require payment of an administrative tax under Article L. 33-1 of the Post and Electronic Communication Code (CPCE).[33]

The use of radio frequencies by French earth stations is covered by authorisations issued by ARCEP. Under the terms of Article L. 42-1 of the CPCE, these authorisations cannot exceed 20 years. ARCEP also imposes a

---

31 Act No. 2004-575 of 21 June 2004 on confidentiality in the digital economy. Available in French only: http://www.legifrance.gouv.fr/affichTexte.do?cidTexte=JORFTEXT000000801164&dateTexte=.

32 Act No. 2004-669 of 9 July 2004 concerning electronic communications and audiovisual services. Available in French only: http://www.legifrance.gouv.fr/affichTexte.do?cidTexte=JORFTEXT000000439399.

33 Code of Post and Electronic Communications. Available in French only: http://www.legifrance.gouv.fr/affichCode.do?cidTexte=LEGITEXT000006070987.

certain number of technical requirements that must be respected by the operators to which the frequencies have been assigned. In addition, operators are required to pay an annual fee to the government for the provision of frequencies and an annual fee for their management.[34] They also have to take the necessary measures to protect the secrecy of private communications as well as the confidentiality of their customers' personal data. Non-compliance with the applicable telecommunication laws and regulations could result in administrative or criminal fines, and also sanctions imposed by ARCEP or other public authorities, including the suspension or withdrawal of the frequency assignment.

Many countries, including most European states, have liberalised their regulatory framework relating to the provision of voice, data and video services. They have also increased the scope for granting authorisations to own and operate earth station equipment and to choose a provider of satellite capacity. Most countries allow authorised providers of communications services to have their own transmission equipment and to purchase satellite capacity without restriction. This facilitates end-user access to Eutelsat's services. Other countries, generally in emerging markets, have maintained strict monopolies. In these countries, a single state entity, generally the public postal, telegraph and telephone authority, often holds a monopoly on the ownership and operation of communications equipment or on the provision of communications or broadcasting services to or from that country, including via satellite. In order to provide services in these countries, Eutelsat may have to negotiate an operating agreement with the state entity, which defines the services to be offered by each party, the contractual terms of the service and tariffs. Depending on national regulatory requirements, operating agreements between Eutelsat and the service provider may require end-user clients to obtain Eutelsat's services through the state entity, with all associated ground services provided by that entity.

[34] Decree No. 2007-1532 of Oct. 24, 2007 concerning charges for the utilisation of radio-electric frequencies payable by holders of licences issued by the regulatory authority for electronic communications and post. Available in French only: http://www.legifrance.gouv.fr/affichTexte.do;jsessionid=FEECB1826080638EC312713980153A93.tpdjo09v_2&dateTexte=?cidTexte=JORFTEXT000000619224&categorieLien=cid and Decision No. 2008-656 of July 8, 2008 extending the licence issued to SA LV & Co. concerning the utilisation of a category D radio service by terrestrial hertzian means with modulating frequency (MFM). Available in French only: http://www.legifrance.gouv.fr/affichTexte.do?cidTexte=JORFTEXT000019415225.

These operating agreements also allow clients to possess and use their own equipment, while requiring them to purchase the Group's services through the state entity. Eutelsat has applied for network operator licences and earth station licences in a certain number of European countries and has been granted a network operator licence in one country.

Despite the liberalisation of national regulations following adoption of the WTO Agreement on Basic Telecommunications Services,[35] some countries require authorisations to operate satellites in-orbit. In these countries, Eutelsat has to obtain authority to provide (i) downlink services from the satellite to the earth station terminals located in these countries ("landing rights") or (ii) uplink services from the earth station terminals to the satellite ("take-off rights"). In certain countries, Eutelsat S.A. has obtained these authorisations for some of its satellites.

As regards access to Eutelsat S.A.'s satellites from the United States, the company has to comply with the regulations of the US Federal Communications Commission (FCC). In 1997, the FCC enacted regulations permitting non-US satellite operators to request access to the US market using non-US satellites, for the provision of both international and domestic services. In 1999, the FCC streamlined the process by creating the "Permitted Space Station List." Where a non-US satellite is added to the FCC's Permitted Space Station List, earth station operators in the United States licensed to operate with US satellites are able to access that specific non-US satellite without additional authorisation from the FCC. Currently two of Eutelsat S.A.'s satellites are included on the Permitted Space Station List. Eutelsat also has to comply with US export control laws and regulations, specifically the Arms Export Control Act,[36] the International Traffic in Arms Regulations,[37] the Export Administration Act[38] and the trade sanctions laws and regulations administered by the US Treasury's Office of Foreign Asset Control. Eutelsat, its service providers, distributors, suppliers and subcontractors that use US technologies (including for communications),

35 Fourth Protocol to the General Agreement on Trade In Services, Apr. 30, 1996, 36 I.L.M. 354, 366. Available: http://www.wto.org/english/tratop_e/serv_e/telecom_e/telecom_e.htm.

36 The Arms Export Control Act, 22 U.S.C. § 2778 (2013). Available: http://www.pmddtc.state.gov/regulations_laws/aeca.html.

37 Amendments to the International Traffic in Arms Regulations, 22 C.F.R. §120.1 (2013). Available: http://www.pmddtc.state.gov/regulations_laws/itar_consolidated.html.

38 Export Administration Act, 15 C.F.R. §730.1 (2013).

export US components for the construction of the Group's satellites or provide launch services outside the United States are all required to obtain permits for the export of technical data and material (under technical assistance agreements) for any material they purchase for the construction of satellites or for satellite launches outside the United States.

Over the last decade, the regulatory environment for satellite communications in the EU has changed drastically. Gradual liberalisation of this sector, as well as the restructuring of EUTELSAT and the privatisation of other international satellite organisations such as INMARSAT and INTELSAT, have created a more open and more competitive market. This trend towards liberalisation has also occurred in a number of other European countries. In particular, countries seeking admission to the EU are adapting their national legislation so as to align it with EU regulations. EU Member States were required to adapt their national regulations by July 2003 to incorporate the provisions of several new EU directives adopted in 2002, which make up the "Telecom Package" concerning access,[39] authorisation,[40] framework,[41] universal service,[42] privacy[43] and radio spectrum.[44] These new regulations apply to electronic communications networks and

39 Parliament and Council Directive 2002/19/EC, access to, and interconnection of, electronic communications networks and associated facilities (Access Directive), 2002 O.J. (L 108). Available: http://eur- lex.europa.eu/LexUriServ/LexUriServ.do?uri=CELEX:32002L0019:EN:HTML.

40 Parliament and Council Directive 2002/20/EC, authorisation of electronic communications networks and services (Authorisation Directive), 2002 (L 108). Available: http://eur-lex.europa.eu/LexUriServ/LexUriServ.do?uri=OJ:L:2002:108:0021:0032:en:PDF.

41 Parliament and Council Directive 2002/21/EC, common regulatory framework for electronic communications networks and services (Framework Directive), 2002 O.J. (L 108). Available: http://eur-lex.europa.eu/LexUriServ/LexUriServ.do?uri=OJ:L:2002:108:0033:0050:EN:PDF.

42 Parliament and Council Directive 2002/22/EC, universal service and users' rights relating to electronic communications networks and services (Universal Service Directive), 2002 O.J. (L 108). Available: *at* http://eur-lex.europa.eu/LexUriServ/LexUriServ.do?uri=OJ:L:2002:108:0051:0051:EN:PDF.

43 Parliament and Council Directive 2002/58/EC, processing of personal data and the protection of privacy in the electronic communications sector (Directive on privacy and electronic communications), 2002 O.J. (L 201). Available: http://eur-lex.europa.eu/LexUriServ/LexUriServ.do?uri=CELEX:32002L0058:EN:HTML.

44 Parliament and Council Decision 2002/676/EC, regulatory framework for radio spectrum policy in the European Community (Radio Spectrum Decision). 2002 O.J. (L 108). Available: http://eur-lex.europa.eu/LexUriServ/LexUriServ.do?uri=OJ:L:2002:108:0001:0001:EN:PDF.

services and generally reduce regulatory requirements in these areas. These directives were transposed into France's national law by the adoption of the above-mentioned July 2004 law (Act 2004-669).

Television broadcasting in the EU is regulated by a Directive of October 1989, known as the "Television without Frontiers" (TVWF) Directive,[45] which has been substantially changed by the June 2007 Directive on the coordination of certain legislative, regulatory or administrative provisions in Member States relating to the performance of television broadcasting activities[46] and the December 2007 "Audiovisual Media Services" (AVMS) Directive.[47] EU Member States were required to implement the 2007 AVMS Directive by amending national laws by December 2009. It was transposed into French law in March 2009.[48] The AVMS Directive provides that each EU Member State shall ensure that television programmes transmitted by broadcasters under its jurisdiction comply with the rules of law applicable to broadcasts intended for the public.

In June 2008, France incorporated into national law[49] its obligations under the 1967 UN Outer Space Treaty[50] and the 1972 UN Liability

45 Parliament and Council Directive 89/522/EEC, co-ordination of certain provisions laid down by law, regulation or administrative action in Member States concerning the provision of audiovisual media services ("Television Without Frontiers" Directive) 1989 O.J. (L 0552). Available: http://eur-lex.europa.eu/LexUriServ/LexUriServ.do?uri=CONSLEG:1989L0552:20071219:EN:PDF.

46 Parliament and Council Regulation Directive 97/36/EC, amending Council Directive 89/552/EEC on the coordination of certain provisions laid down by law, regulation or administrative action in Member States concerning the pursuit of television broadcasting activities, 1997 O.J. (L 202). Available: http://eur-lex.europa.eu/LexUriServ/LexUriServ.do?uri=celex:31997L0036:en:html.

47 Parliament and Council Directive 2007/65/EC, Amending Council Directive 89/552/EEC on the coordination of certain provisions laid down by law, regulation or administrative action in Member States concerning the pursuit of television broadcasting activities, 2007 O.J. (L 332). Available: http://eur-lex.europa.eu/LexUriServ/LexUriServ.do?uri=CELEX:32007L0065:en:NOT.

48 Act No. 2009-258 Mar. 5, 2009 concerning audiovisual communication and a new public television service. Available in French only: http://legifrance.gouv.fr/affichTexte.do?cidTexte=JORFTEXT000020352071&categorieLien=id.

49 Act No. 2008-518 of June 3, 2008 concerning space operations. Available in French only: http://www.legifrance.gouv.fr/affichTexte.do?cidTexte=LEGITEXT000018939303.

50 Treaty on Principles Governing the Activities of States in the Exploration and Use of Outer Space, Including the Moon and Other Celestial Bodies art. 6, Jan. 27, 1967, 18 U.S.T. 2410, 610 U.N.T.S. 2050. Available: http://www.oosa.unvienna.org/pdf/publications/STSPACE11E.pdf.

Convention.[51] Two decrees were published in June 2009, one relating to control of premises[52] and the other relating to authorisations,[53] which is of major importance to Eutelsat S.A. The entire legal framework resulting from the 2008 Space Operations Act entered into force in December 2010 and the main provisions of the law are summarised as follows:

- an authorisation regime for space operations that may incur France's international liability, such as the launch of a space object from France and, for a French operator, the launch of a space object from abroad, the control of a space object in outer space or the transfer of control of a space object that has already been authorised. Contrary to the expectations of Eutelsat S.A., no global licence may be granted for launches, either from French or non-French territory and authorisation will be necessary for each individual launch;
- a licence regime for satellite operators. Eutelsat has been granted a global licence for operating space objects;
- an insurance obligation which covers the whole space operation of a satellite (alternatively other financial guarantees may be considered); and
- a State guarantee in the case of damage, except for wilful damage, under which, should an operator be sentenced to pay compensation to a third party, France would be bound to pay the excess of the insurance amount. This guarantee would apply for Eutelsat S.A. in case of damage resulting from a launch from French territory or for damage caused on the earth's surface or to an aircraft in flight during the in-orbit phase. In practice it would seem that a large part of Eutelsat S.A.'s activity would be excluded from this benefit provided by the law.

### 6.6. *Activities of EUTELSAT S.A.*

Eutelsat S.A, which is subject to income tax in the same way as any company incorporated in France, derives its revenues principally by leasing

---

[51] Convention on international liability for damage caused by space objects (Liability Convention) 29 Mar. 1972, 24 U.S.T. 2389, 961 U.N.T.S. 187. Available: http://www.oosa.unvienna.org/pdf/publications/STSPACE11E.pdf.

[52] Decree No. 2009-640 of June 9, 2009 on the application of the conditions of chapter VII of Act No. 2008-518 of 3 June 2008 concerning space operations. Available in French only: http://textes.droit.org/JORF/2009/06/10/0132/0001/.

[53] Decree No. 2009-643 of June 9, 2009 concerning licences issued in accordance with Act No. 2008-518 of 3 June 2008 concerning space operations. Available in French only: http://textes.droit.org/JORF/2009/06/10/0132/0030/.

satellite transponder capacity. The company leases satellite capacity and provides other services both to distributors (which re-lease the capacity to third party users, which, in turn, provide their services to end-users) and to customers (which use Eutelsat S.A.'s capacity themselves). In accordance with transmission plans and traffic information, the satellite capacity is used for video application services (68.6 percent of revenues in the financial year ended 30 June 2011),[54] data and value added services (20.4 percent of revenues in the financial year ended 30 June 2011) and multi-usage capacity (11 percent of revenues in the financial year ended 30 June 2011), which includes capacity leased to governments and administrations.

Eutelsat's pricing policy changed significantly as a result of the restructuring. The company calculates its tariffs to respond to market conditions. However, the majority of Eutelsat S.A.'s existing lease agreements were made prior to the restructuring. As is standard in the industry, Eutelsat S.A.'s tariffs for lease capacity depend on a number of factors including (1) the orbital location of the satellite, (2) the number of antennae accessing the satellite, (3) the geographic region covered by the satellite, (4) the volume and type of application that the satellite can service, (5) the type and duration of the lease capacity agreement, (6) the type and number of transponders, the amount of bandwidth and the power of the satellite to be used, (7) the rates charged by competitors for similar capacity and services, and (8) whether the service is pre-emptible (that is, whether Eutelsat S.A. retains the right to use the lease capacity for another customer in case of the failure or malfunctioning of transponders leased by that other customer). Eutelsat S.A.'s tariff strategy allows it to create a specifically tailored structure for an existing or planned satellite to respond to customer demand in certain markets or to adopt a competitive tariff strategy in order to access new geographical markets. Customer contracts made before restructuring and transferred to it as part of EUTELSAT's assets remained subject to their original pricing and payment terms. The majority of these provide capacity at a fixed price agreed at the date of the contract and applicable for the entire operational life of the relevant satellite. Contracts have never contained most-favoured customer conditions, either before or after the restructuring. Since restructuring, Eutelsat S.A. has revised the

---

54 "Eutelsat Powers Up A New Orbital Position To Drive Expansion In High Growth Markets," Eutelsat Communications S.A. *Press release 41/11*, July 28, 2011, accessed February 28, 2013. Available: http://www.eutelsat.com/news/compress/en/2011/pdf/PR%204211%20Eutelsat%203B.pdf.

terms and conditions of its standard contracts for the provision of space segment capacity and has set up a non-exclusive distribution agreement. The company's standard terms and conditions provide for various reservation options, as well as for the firm leasing of satellite capacity on its satellites. Pursuant to these contracts, a customer may procure capacity services from Eutelsat S.A. through earth stations or other ground segment facilities and use any existing or new earth station to distribute services using Eutelsat S.A.'s capacity. In accordance with Eutelsat S.A.'s standard lease of capacity agreements, customers must obtain operating licenses from the relevant regulatory authority, secure rights to maintain earth stations and comply with Eutelsat S.A's technical specifications. In addition, to ensure security of payment of the leased capacity and of the customer's observance of its obligations, Eutelsat S.A. may require that a customer provide a bank guarantee or other appropriate security.

During the 20 years in which EUTELSAT operated as an intergovernmental organisation, it developed strong relationships with large telecommunications service providers in over 50 countries, which have been maintained by Eutelsat S.A. Its distribution network strives to maximise its fill factor and associated backlog objectives by means of long term agreements with its large distributors for the lease of new capacity and services. Capitalising on the commercial network which it inherited from EUTELSAT, Eutelsat S.A. proposed to each of the former Signatories that they become a distributor of its satellite capacity and services on a non-exclusive basis. To qualify as a distributor, the distributor's business activities must be dedicated mainly to the supply of value added and/or packaged communication services for the benefits of its customers. The distributor must also be in a position to bundle its services with its own ground communication infrastructures (earth stations) and to allocate skilled staff, marketing and commercial services to achieve maximum sales of Eutelsat S.A.'s services, subject always to the requirements of its own customers. The distributor is also required to refrain from granting preferential treatment and/or exclusive rights to any of Eutelsat S.A.'s competitors. Distributors are required to report regularly on market trends and Eutelsat S.A.'s service performance and positioning, to provide Eutelsat S.A. with forecasts and other relevant marketing information and to engage in marketing activities in agreement with Eutelsat S.A. Under Eutelsat S.A.'s distribution agreements, the company may not engage in direct sales to users currently serviced by its distributors, unless the user solicits its services. Eutelsat S.A.'s distribution agreements provide for discounts to its distributors and are initially for one year. Upon expiration of this initial term, they are automatically renewed for successive one year periods unless terminated by either party giving notice 30 days before the expiration of the initial term or any successive term. The main advantage of a distribution network for Eutelsat S.A. is the limitation of its commercial exposure, as it is the distributors who bear any financial risks associated with the marketing of capacity in new geographical markets or marketing of new applications such as broadband multimedia services.

On 2 July 2001, all intellectual property developed by EUTELSAT, consisting mainly of trademarks, was transferred to Eutelsat S.A., which is now the owner thereof. There is a special agreement between EUTELSAT IGO and Eutelsat S.A. for the use of the trademark EUTELSAT provided that it is written in capital letters and followed by IGO (for intergovernmental organisation) while the companies of the Eutelsat Group write Eutelsat in small letters in order to avoid potential confusion. To date the Eutelsat Group owns 26 patent families, 2 of which are held on a co-ownership basis with ESA. Research and development activities are mainly directed at the multimedia business.

Since 2001, Eutelsat S.A. has launched 16 satellites and 6 more satellites are to be launched by mid 2013, representing an investment of 2.5 billion euros in satellites and launchers since restructuring.

Eutelsat S.A. has an extensive insurance programme covering both phases of a satellite's lifespan, i.e. launch (launch insurance policy) and orbit (in-orbit lifetime insurance policy). It also has on-ground and in-orbit third-party liability insurance and other customary commercial insurance policies regarding its operations. Its launch and in-orbit lifetime insurance policies include provisions and exclusions that are customary in space insurance, including those relating to loss or damage caused by *force majeure* events (such as armed conflict), natural occurrences related to a satellite's environment in space, intentional acts by Eutelsat S.A. and damage to third parties. The company also has insurance to cover risk of non-payment by its customers and third-party liability insurance covering its Corporate Officers, Directors and senior managers, as well as the senior managers of its subsidiaries, in the performance of their duties. Eutelsat has also taken out standard third-party liability insurance covering its ground operations, comprehensive insurance for its on-ground equipment and various assistance policies for its employees and "guests."

## 7. *EUTELSAT IGO Since Resructuring*

### 7.1. *Relations between EUTELSAT IGO and EUTELSAT S.A.*

The relationship between Eutelsat S.A. and EUTELSAT IGO is governed by an agreement (the "Arrangement"[55]) which came into force on 2 July 2001 and has subsequently been amended four times. The Arrangement states that, on the understanding that the management of Eutelsat S.A. is carried out on a sound economic and financial basis, Eutelsat S.A.'s principal obligation shall be to observe the four Basic Principles listed in Article III(a) of the Amended EUTELSAT Convention. It also provides a mechanism for settling disputes, including by arbitration.

Under the terms of the Arrangement, Eutelsat S.A. has to respect certain obligations to EUTELSAT IGO. It has to give 60 days' notice of any proposal to change its Articles of Association where these might materially affect its observance of the Basic Principles. The company has to inform EUTELSAT IGO, and take into account any recommendation made by EUTELSAT IGO, in the event of any major changes to its operating, technical, marketing or financial policies that might materially affect the observance of the Basic Principles. It must obtain written prior approval from EUTELSAT IGO if it intends to go into voluntary liquidation, or if it intends to merge or combine with another entity. Subject to certain conditions, it has to name EUTELSAT IGO's Executive Secretary a Censeur (Non-voting Director) on its Board of Directors and, last but not least, it has to finance EUTELSAT IGO's annual operating costs. The Arrangement also states EUTELSAT IGO's commitments to Eutelsat S.A. The organisation must make every effort to ensure that Eutelsat S.A. can make use of all frequency assignments acquired or filed with the ITU Radiocommunication Bureau as of 2 July 2001. Any proposed amendment to the Amended EUTELSAT Convention that is liable to affect EUTELSAT IGO's performance of its activities must be submitted to Eutelsat S.A., which shall have six weeks in which to reply with its observations.

A Joint Commission made up of representatives of EUTELSAT IGO and Eutelsat S.A. meets at least once per year to ensure that Eutelsat S.A. is observing the Basic Principles. Eutelsat S.A. sends EUTELSAT IGO extracts

[55] The *Arrangement between the European Telecommunications Satellite Organisation and Eutelsat S.A.* is not publicly available.

from its Five-Year Strategic Plan and its certified annual accounts and examines with EUTELSAT IGO the impact on its activity or on its observance of the Basic Principles caused by any changes in applicable regulations, particularly European or French.

In his capacity as Censeur, the Executive Secretary of EUTELSAT IGO has access to information under the same conditions as those which apply to a Director and he is able to attend, but not vote at, meetings of Eutelsat S.A.'s Board of Directors.

### 7.2. *Structure of EUTELSAT IGO*

EUTELSAT IGO, since the restructuring, consists of the Assembly of Parties and a Secretariat, which is the permanent organ of the organisation, headed by the Executive Secretary. The Assembly of Parties meets ordinarily every two years and is assisted by an Advisory Committee, which meets at least every six months. The functions of the Assembly of Parties, as stated in Article IX of the Amended EUTELSAT Convention, are to review the activities of Eutelsat S.A. which relate to the Basic Principles and to ensure observance by Eutelsat S.A. of the Basic Principles in accordance with the Arrangement. Additionally the Assembly of Parties takes decisions on proposed changes to the Arrangement (such changes being subject to mutual agreement of the parties to the Arrangement), and any proposal to terminate or amend the Convention (including submitting to the company any proposed amendment that is liable to affect the performance of its activities). It also takes appropriate decisions in order to ensure continuity regarding rights and obligations under international law, in particular under the ITU Radio Regulations for the use of frequencies deriving from the operation of EUTELSAT space segment transferred to the company Eutelsat S.A. The Assembly of Parties decides on accession or withdrawal of a Party, on the appointment or removal from office of the Executive Secretary, on staff employment conditions and adopts a budget for the Secretariat. It also would have to approve any change in the location of Eutelsat S.A. and consider any complaints submitted to it by Parties.

The Advisory Committee, composed of representatives of six Parties, was set up by the Assembly of Parties in October 2000 in order to give advice to the Executive Secretary on budgetary and staffing issues, in particular during the establishment phase of the EUTELSAT IGO Secretariat under the new legal structure after restructuring. The mandate adopted by the Assembly of Parties states that the Committee shall consult with the Executive Secretary and give its guidance and advice on Secretariat budget, accounting and auditing procedures, staffing structure and employment terms, preparation of Joint Commission meetings, if required, and any other matter delegated to it by the Assembly of Parties. The Advisory Committee produces reports to the Parties on the results of its work, and also submits a report to the Assembly of Parties. In 2009, the Assembly of Parties decided that a maximum of two representatives of Parties may attend meetings of the Advisory Committee as observers.

The Executive Secretary is the legal representative of EUTELSAT IGO and is elected by the Assembly of Parties. His/her mandate is of four years and s/he is directly responsible to the Assembly of Parties for the performance of all functions of the EUTELSAT IGO Secretariat. The Executive Secretary of EUTELSAT IGO acts as the focal point between EUTELSAT IGO and Eutelsat S.A and represents the organisation in the Joint Commission. In monitoring the activities of Eutelsat S.A. so as to ensure that the company observes the four Basic Principles, the Executive Secretary regularly analyses the developments of the Eutelsat Group, so that the organisation can respond swiftly in case major changes occur either in the company's activities or in its regulatory environment, which could affect the respect of the Basic Principles. This includes, *inter alia*, following regulatory developments (at the national, European and international level), interpretation of the basic legal texts of the organisation and preparation of legal opinions. The Executive Secretary represents the organisation in bilateral and multilateral meetings with entities and other international organisations and ensures the role of interface between the ITU Radiocommunication Bureau, the Notifying Administration for EUTELSAT satellite networks and Eutelsat S.A. for radio frequency issues in accordance with Article III(b) of the Amended EUTELSAT Convention. The Executive Secretary prepares and proposes budgets for the organisation for adoption by the Assembly of Parties and presents the organisation's accounts in conformity with International Financial Reporting Standards, producing an annual report in accordance with Article XX of the Amended Convention.

### 7.3. *Letter-Agreements*

Since 2001, it has been necessary to introduce new instruments to ensure that EUTELSAT IGO is able to fulfil its role. A Letter-Agreement was signed by the Executive Secretary and the major shareholders of Eutelsat S.A. in December 2004. In June 2004, the Chairman of the Management Board of Eutelsat S.A wrote to inform EUTELSAT IGO about the proposal of the Supervisory Board of the company to change the Articles of Association in relation to the principle of an IPO for the company. The Executive Secretary at the time considered that these proposed changes were of such importance that an extraordinary meeting of the Assembly of Parties was convened in July 2004 to allow Parties to review the matter and decide how to proceed. A legal opinion on the interpretation of the information provided by the company and its potential long term impact was prepared. The Assembly of Parties requested the Executive Secretary to enter into negotiations with Eutelsat S.A. to define new ways of supervising the company so as to guarantee its observance of the Basic Principles and the principles underlying the restructuring of EUTELSAT (in particular the protection of the rights of small shareholders and the preservation of the company's European identity), together with the various commitments entered into by the former Signatories (who became the original shareholders of the company) regarding pre-emption rights for share transactions and the company's IPO, before any amendments to the Articles of Association were brought into effect. From August to December 2004 negotiations between EUTELSAT IGO and Eutelsat S.A. took place and resulted in the amendment of the Arrangement between the organisation and the company as well as the above mentioned Letter-Agreement. The main points of the agreement consisted in a change in the company's Articles of Association to introduce a "collège de censeurs" of two Observers in the Board of Directors, one of these being assigned to the organisation and held by the Executive Secretary in person and the assignment of two seats on the Board of Directors to shareholders holding less than seven percent each. The amendment to the Arrangement consisted of including the rights of supervision and of the Observer (Censeur) role of the organisation, with a description of the principles underlying the exercise of this role, in particular the procedures for preparatory sessions prior to meetings of the Board of Directors and for the distribution of information about the company's activities to Parties, including mediation procedures, should there be a difference of opinion over this matter. The amendment also covered the enlargement of the role of the Joint Commission to hold a preparatory meeting before each meeting of the Board of Directors and to consider on an annual basis the Five Year Strategic Plan of the company, its financial results, all regulatory issues related to the satellite telecommunications sector, including ITU and EU regulations as well as French legislation applicable to the company in the context of the Basic Principle of public service/ universal service obligations and to examine matters related to the further definition of the Basic Principles.

Eutelsat Communications S.A. and EUTELSAT IGO signed another Letter-Agreement in September 2005, which came into force on 6 December 2005, following the IPO. In this, Eutelsat Communications S.A. undertook to give EUTELSAT IGO's Executive Secretary a seat as Censeur on the Board of Directors of Eutelsat Communications, to ensure that Eutelsat S.A. was at all times able to honour its undertakings made pursuant to the Arrangement and not to take any decision which might entail any breach of the said undertakings by Eutelsat S.A. Eutelsat Communications S.A. also agreed to inform the Executive Secretary, in his/her capacity as non-voting Director, of any decision taken by Eutelsat Communications which might affect Eutelsat S.A.'s compliance with the Basic Principles and to communicate to him/her all useful information on such matters. Furthermore, it would inform EUTELSAT IGO of any crossing of a legal threshold or of a threshold contained in the Articles of Association, which has been notified to it by a shareholder. Eutelsat Communications S.A. undertook not to propose or vote for any proposal that Eutelsat S.A. distribute dividends in excess of the amount of its annual net income and/or annual net income plus retained earnings and/or which would result in Eutelsat S.A.'s net debt/EBITDA ratio rising above 3.75/1, given that this ratio would not be considered as having been exceeded where any excess comes as a result of any external growth operation and that the notion of dividends was that defined under Article L. 232-12 of the French Commercial Code.[56] It was agreed that any undertakings given by Eutelsat Communications, in particular in relation to its financial needs, present or future, would not in any way result in cross default by Eutelsat S.A., unless such undertakings given by Eutelsat Communications were also given in Eutelsat S.A.'s direct interest. The consolidated Group debt would be maintained at a level that was not contrary to market practice and sound management of the Eutelsat Group. A minimum amount of equity would be maintained in Eutelsat S.A. in compliance with sound financial management of Eutelsat S.A., allowing it to continue to comply with the Basic

---

[56] French Republic, *Commercial Code*. Available: http://www.legifrance.gouv.fr/html/codes_traduits/commercetextA.htm.

Principles. The Letter-Agreement also provided for the creation of a Co-ordination Committee, whose main tasks were (i) to exchange useful information and views for the proper implementation of the Letter-Agreement, (ii) to examine any request for the removal of confidentiality restrictions on information received by the Executive Secretary in his role of Censeur, and (iii) to examine in particular the annual accounts and the list of third-party experts designated to resolve any problem arising as to what information may be circulated by the Executive Secretary to the Parties to the Convention. The Letter-Agreement would become null and void upon the expiry of the Arrangement pursuant to its terms and conditions. (It should be noted that the Arrangement may only be terminated by mutual agreement). EUTELSAT IGO and Eutelsat Communications could, however, terminate or amend the Letter-Agreement at any time upon mutual agreement, in particular in the event where such termination or amendment proved to be helpful in facilitating the development of the Group. In the event of assignment of Eutelsat S.A. shares by Eutelsat Communications S.A., the latter shall inform the proposed transferee of the content of the Letter-Agreement, it being understood that Eutelsat Communications S.A. remains bound, in any event, by its undertakings until the expiry of the Letter-Agreement. A mechanism for settling disputes by arbitration was also included.

Since 2004, in accordance with the provisions made in the Arrangement and subsequently in the above-mentioned Letter-Agreements, the Executive Secretary of EUTELSAT IGO has performed the duties of Censeur (observer) on the Board of Directors of Eutelsat Communications and Eutelsat S.A. and has participated in all the meetings of the Board of Directors, to date 57 meetings of the Board of the operating company Eutelsat S.A. and 67 Board meetings of the holding company Eutelsat Communications S.A. As a result of this, the Executive Secretary has been able to provide the Assembly of Parties and the Advisory Committee with reports and information to enable them to review the latest developments and to consider whether Eutelsat S.A. fulfils its obligations to the organisation and is in a position to continue to observe the Basic Principles. The Assembly of Parties has recognised that the role of Censeur on the Boards of Directors of Eutelsat S.A. and Eutelsat Communications provides the Executive Secretary with a practical means of monitoring the company with respect to its ability to observe the Basic Principles, subject to him/her observing the rules of confidentiality stated in the Arrangement.

### 7.4. *Activities of EUTELSAT IGO*

In addition to the role of overseeing Eutelsat S.A., EUTELSAT IGO has maintained its role in ensuring the continuity of international collectively-owned rights for frequencies and orbital locations and has adapted to the changing environment in the satellite industry and considered the market and future prospects for the fixed and mobile satellite operating companies, other intergovernmental satellite organisations and international entities within the same field of activity.

An example of the activities undertaken by EUTELSAT IGO in the maintenance of the continuity of frequency rights was its role in obtaining the support of Parties over the question of deliberate jamming of programmes broadcast by Eutelsat S.A. via radio frequencies held by EUTELSAT IGO, as described previously. The Executive Secretary wrote to the Parties informing them about this and requesting their co-operation, stressing that it could be significantly harmful to Eutelsat S.A., over which EUTELSAT IGO has a supervisory role. Deliberate jamming could affect the credibility in general of satellites by posing a threat to the secure transmission of programmes by satellite. Since May 2009, repeated interference had been experienced on several radio and television channels broadcast via Eutelsat satellites. Complaints had been lodged with both the ITU and the administration at the origin of the interference, but without success. In early February 2010, the jamming increased to the level where it became persistent. EUTELSAT IGO forwarded information about the interference to the ANFR which, acting as Notifying Administration for EUTELSAT IGO, officially notified the ITU Radiocommunications Bureau. In accordance with the ITU Radio Regulations, EUTELSAT IGO, via the ANFR, formally requested that this matter be considered at the next meeting of the ITU Radio Regulations Board in March 2010. At the request of the Executive Secretary, association letters were received from 28 Member States of EUTELSAT IGO to support this approach. The Executive Secretary also brought the subject of deliberate jamming to the attention of the Legal Subcommittee of UNCOPUOS[57] in 2010, where he presented a report on

[57] United Nations Committee on the Peaceful Uses of Outer Space. Available: http://www.oosa.unvienna.org/oosa/COPUOS/copuos.html.

the situation and appealed to the State in question to cease such jamming in conformity with the decision taken by the ITU Radio Regulations Board.[58]

Another area in which EUTELSAT IGO has been active is in obtaining the inversion of the criteria stated in the TVWF Directive which determines which State has jurisdiction over a television channel established outside the EU. Eutelsat S.A. had found itself in a position where it was transmitting programmes originating from a country outside the EU and received by the general public within EU territory which were in breach of public order. According to the original TVWF Directive, France was responsible for ensuring that Eutelsat ceased transmission of these programmes, since the company was established in France, and, in 2005, the French Broadcasting Authority (CSA) implemented this Directive and issued formal notice to Eutelsat to stop broadcasting the offending channels. The Assembly of Parties considered the question and in particular the risk that Eutelsat S.A. might be put in a disadvantageous position in relation to other satellite operators providing coverage to EU countries, which operated from countries to which the TVWF Directive either did not apply or in which it was not fully implemented. Any television channel prohibited from transmitting via the Eutelsat system could still continue to transmit into the EU via other satellite systems with European coverage, if they were under the authority of a country which had not implemented the TVWF Directive. Eutelsat S.A. requested the organisation to support the initiatives taken at EU level to invert the criteria of the TVWF Directive, with the aim of making the country where the uplink station is situated the principal criterion used to decide which State had authority over the television channel instead of the State of the satellite operator. The Assembly of Parties passed a resolution urging mutual recognition by all EU States of a decision by one of them to prohibit the broadcasting of a television channel originating from a country outside the EU on grounds of breach of public order, such as incitement to violence or hatred for reasons of race, sex, religion or nationality. The combined initiatives of EUTELSAT IGO and its Members States together with Eutelsat S.A. led to the inversion of the criteria, embodied in a new AVMS Directive, which updated the former TVWF Directive and was transposed into French Law in March 2009.

58 "ITU Radio Regulations Board urges Iran to end interference hampering EUTELSAT satellite operations," International Telecommunication Union press release, Mar. 26, 2010, accessed Feb. 28, 2013. Available: http://www.itu.int/newsroom/press_releases/2010/14.html.

EUTELSAT IGO has also given extensive consideration to the development of a new national regime for satellite operators in France. In 2007, the Advisory Committee of EUTELSAT received a presentation from Eutelsat S.A. about a draft French law on space activities which was to be reviewed by the French Parliament in the near future. The Executive Secretary shared the serious reservations expressed by the company about this legislative text which could potentially have had harmful consequences for Eutelsat S.A.'s satellite operations, in particular in the areas of launch, maintenance of position, de-orbiting and handover to a third party. In view of these concerns, the Committee requested that this issue be considered by the Assembly of Parties, with Eutelsat S.A. invited to make a presentation. While neither EUTELSAT IGO nor Eutelsat S.A. questioned the legitimacy of a new legal national framework of this type, it was appropriate for EUTELSAT IGO to consider the potential impact of this new national legislation on Eutelsat S.A., in particular in relation to its ability to fulfil its obligations to the organisation. The French Space Operations Act, published in June 2008, is far reaching and highly comprehensive (the Prime Minister at the time of the review by the Council of State suggested that the French legal framework might in the future become a standard of reference for possible European legislation) and cannot be considered in the same way as any ordinary new law, such as those of a financial, social or commercial nature, which apply to all companies established and operating under French jurisdiction, since, in practice, Eutelsat is virtually the only satellite operator to which the law currently applies. In April 2009, the Assembly of Parties requested Eutelsat S.A. and the Party of France to give specific attention to the obligations incumbent upon them to ensure the fulfilment of the commitments made at the time of restructuring and stated in the Amended EUTELSAT Convention. The company received considerable support from the organisation and its representatives, and, with the assistance of the Party of France, Eutelsat's concerns were taken into consideration during the drafting process of the Act and its related technical regulations. The technical regulations for the implementation of the Act were agreed in late 2009 in collaboration with the French National Centre for Space Studies (CNES) and Eutelsat S.A. informed the Executive Secretary that it did not foresee any problems resulting from these. However, as requested by the Parties, the Executive Secretary will continue to consider whether it has any impact on Eutelsat S.A.'s ability to observe the Basic Principles one year after the full entry into force of the new French Space legal framework which took place in December 2010.

Mindful of the similarity of some of the activities undertaken by EUTELSAT IGO, IMSO (as the International Mobile Satellite Organisation is known since its restructuring) and ITSO (as the International Telecommunications Satellite Organisation is known since its restructuring), the Heads of these three organisations have, since January 2008, held tripartite meetings, where they discuss activities of mutual interest. A Tripartite Memorandum of Understanding was signed in December 2009, following approval by each of the organisations' Assemblies. It was agreed that the three organisations would explore together the possibilities of a joint approach to initiatives for instance in relation to harmonising the regulatory framework applicable to satellite communications to facilitate market access, capacity building, training, and suchlike. All three organisations have stressed the importance of raising awareness of satellites as a tool to implement a broadband infrastructure. This ongoing co-operation resulted in a meeting in February 2010 between the Heads of the three organisations and the ITU Secretary-General to discuss how the organisations could collectively assist in the target of the World Summit on the Information Society (WSIS) in relation to "connecting the unconnected by 2015." This meeting was followed by an informal meeting with the Director of the UN Office of Outer Space Affairs. Both welcomed the involvement of the satellite IGOs, saying it was pertinent, useful and timely. It is planned to continue these tripartite meetings as recent developments have proved that the discussion of matters of mutual interest is fruitful at a wider international level.

EUTELSAT IGO also plays a role in the promotion of satellite solutions to address the digital divide. The Executive Secretary is a Founding Commissioner of the Broadband Commission for Digital Development[59] which was established in 2010. He participates in its activities which are scheduled to continue until 2015. Set up at the initiative of the ITU Secretary General and the Director General of UNESCO, the Broadband Commission for Digital Development comprises leaders from industry, civil society, government representatives, UN agencies and the creative sphere. Its aim is to accelerate the achievement of the UN Millennium Development Goals.[60] The Executive Secretary has to date participated in three meetings

59 Broadband Commission for Digital Development. Available: http://www.broadband commission.org/.

60 "United Nations Millennium Development Goals," United Nations, accessed Feb. 28, 2013. Available: http://www.un.org/millenniumgoals/.

of the Broadband Commission and in some of the working groups, where he has, on behalf of IMSO, ITSO and EUTELSAT IGO, stressed that satellite technology should be seriously considered as a valuable means for providing an immediate, ubiquitous and cost-effective connectivity to internet for all. Commissioners from the developing countries have been particularly interested in the use of satellite technology.

Since June 2008, EUTELSAT IGO has held permanent observer status at meetings of UNCOPUOS and the Executive Secretary regularly participates in the meetings of the Technical and Scientific Subcommittee, the Legal Subcommittee and the plenary meetings of UNCOPUOS. In June 2009, UNCOPUOS agreed upon a work programme about long term sustainability of space operations, which was then endorsed by the UN General Assembly in Resolution 64/86.[61] The Executive Secretary is actively involved in the working group established to consider this issue, which is of great importance since space debris and pollution in outer space is a major international concern. EUTELSAT IGO has also brought the subject to the attention of its Member States.

## 8. *Conclusion*

In 2 July 2001, when EUTELSAT was restructured, there were questions about how the relationship between the new operating company, Eutelsat S.A., and the restructured intergovernmental organisation, EUTELSAT IGO, was really going to work in practice. Some had doubts about the ongoing role of the organisation and questioned whether it would continue to be of relevance. There was also some apprehension about whether the company could reach its full potential while maintaining its obligations to EUTELSAT IGO. Time has shown that these initial concerns were unfounded, despite various ups and downs, metamorphoses and misadventures experienced by Eutelsat S.A. and long and difficult negotiations on several occasions between the company, its shareholders and EUTELSAT IGO.

EUTELSAT IGO's role has proved to be extremely relevant and Eutelsat S.A., along with the Eutelsat Group as a whole, has been highly successful in a way that would not have been possible if its activities had remained

[61] G.A. Res. 64/86, U.N. Doc. A/RES/64/86 (Jan. 13, 2010). Available: http://www.oosa.unvienna.org/pdf/gares/A_RES_64_086E.pdf.

within an intergovernmental organisation, due to fundamental changes in the economic, commercial, competitive and regulatory environment. The decisions taken at the time of restructuring have proved to be right, in particular the decision to amend the original EUTELSAT Convention to accommodate the change of structure and to enshrine the provisions and principles of the Amended EUTELSAT Convention into the Arrangement signed between Eutelsat S.A. and EUTELSAT IGO. Over time, it has been necessary to complete these instruments with additional agreements, such as the Letter-Agreements negotiated with Eutelsat S.A. and its shareholders. Both EUTELSAT IGO and Eutelsat S.A. face the future with confidence.

# Chapter Five

## The Role and Function of Residual International Intergovernmental Satellite Organisations Following Privatisation

*Maury J. Mechanick*

### 1. *Introduction*

The move to privatise each of the three major international satellite organisations (ISO or ISOs) – INTELSAT, INMARSAT and EUTELSAT[1] – raised a number of issues regarding whether and how their respective public services missions should be preserved following privatisation. This chapter will consider how these public policy issues were addressed throughout the privatisation process for each of the ISOs. In particular, consideration will be given to the rationale for the preservation of some form of

[1] The proper name of each of the international organisations at the time of privatisation was the International Telecommunications Satellite Organisation, the International Mobile Satellite Organisation (which, prior to a formal name change in 1994, had originally been named the International Maritime Satellite Organisation), and the European Telecommunications Satellite Organisation. Prior to privatisation, these organisations were known by the acronyms INTELSAT, INMARSAT and EUTELSAT respectively. Following privatisation, while the names of the international organisations remained the same, the acronyms used were changed because the prior acronyms were used by the privatised companies that were created through the privatisation process. As such, INTELSAT became known as ITSO, INMARSAT became known as IMSO, and EUTELSAT became known as EUTELSAT IGO. To avoid any confusion herein, when reference is made to the acronym for each of those international organisations, whether in the form it took prior to or following privatisation, that name will appear in all capital letters (INTELSAT or ITSO, INMARSAT or IMSO, and EUTELSAT or EUTELSAT IGO). When reference is made to the name of each of the privatised companies that was created through the privatisation process, an initial capitalization format will be used (Intelsat, Inmarsat and Eutelsat).

intergovernmental oversight over the privatised satellite companies that emerged from the privatisation process; the scope of authority that was to be conferred on those oversight bodies, including their Core Principles or primary purposes; the anticipated duration of their continued existence and mechanisms for their funding; their role and responsibilities – if any – with respect to the orbital resources previously held by the ISOs; the manner in which the public service obligations were to be discharged; and various internal governance issues. This will be followed by an examination of specific issues that have arisen (or may arise in the future) following privatisation and how these oversight bodies have dealt (or are likely to deal) with them, as well as the relationship between these oversight bodies and other international organisations. This chapter concludes with an overall assessment of the functioning and success to date of these oversight bodies.

As more than a decade has now passed since the implementation of the decision to privatise each of the ISOs, the time is certainly ripe for a review of the decision-making processes that occurred, including the underlying forces at play, as well as the outcomes that were achieved, particularly with respect to the decision to retain some form of intergovernmental oversight of the privatised commercial entities. The perspectives gained from the passage of time since those decisions were taken further contribute to this undertaking, allowing for a more detached assessment of those processes and outcomes as well as for a critical reassessment of those decisions in light of subsequent developments.

### 1.1. *Overview of the Privatisation Decision-Making Process*

As originally established, each of the ISOs possessed a dual character in which the fulfilment of certain governmental public policy objectives and commercial interests were blended together into a single entity. In effect, each ISO simultaneously functioned as a treaty-based organisation comprised of governments (referred to as Parties) discharging a policy-making role and as a facilities-based operator owning and operating a satellite system on which commercial services were provided.[2] Funding for the

[2] The underlying treaty structure for each of the ISOs consisted of two separate agreements, one of which focused on the intergovernmental aspects of the organisation and the other which focused on the operational aspects of the organisation. The intergovernmental agreements, which were was signed by governments or Parties thereto, were known as the

commercial operations of the organisations and managerial oversight of those operations was not provided by the Parties but rather was provided by designated operating entities (referred to as Signatories) from each of the countries that were members of the international organisation.[3]

The ultimate decision makers bearing primary responsibility for the decision to privatise were the Parties, although the Signatories and the professional management of each of the organisations also contributed to this deliberative process, supplying many of the specific details as to how privatisation would occur. Other interested players, including competitors and, in the case of the United States, the US Congress, also played a major role in influencing aspects of the privatisation process that unfolded. Furthermore, while the deliberative processes undertaken by each of the three organisations were conducted separately, the governments involved clearly overlapped, as in fact did many of the specific individuals involved in the

INTELSAT Agreement, the INMARSAT Convention or the EUTELSAT Convention. See for INTELSAT the Agreement relating to the International Telecommunications Satellite Organisation (INTELSAT), done 20 Aug. 1971, entered into force 12 Feb. 1973, as amended 17 Nov. 2000, amended version applied provisionally 18 July 2001, entered into force 30 Nov. 2004, 1220 U.N.T.S. 21; for INMARSAT the Convention on the International Maritime Satellite Organisation (INMARSAT), done 3 Sept. 1976, entered into force 16 July 1979, as amended 1998, amended version applied provisionally 15 April 1999, entered into force 31 July 2001, 1143 U.N.T.S. 105; and for EUTELSAT the Convention establishing the European Telecommunications Satellite Organisation (EUTELSAT), done 15 July 1982, entered into force 1 Sept. 1985, as amended 20 May 1999, amended version applied provisionally 2 July 2001, entered into force 28 Nov. 2002, 1519 U.N.T.S. 175. For purposes of citation to particular Articles of each of those agreements, the following terminology will be used. In the case of references to the version of the treaty documents in force prior to privatisation, those agreements are referred to as the INTELSAT Agreement, INMARSAT Convention and EUTELSAT Convention. In the case of references to the version of the treaty documents in force following privatisation, those agreements are referred to as the ITSO Agreement, IMSO Convention and EUTELSAT IGO Amended Convention. With specific regard to the IMSO Convention, which was further amended subsequent to privatisation, the citation provided will refer, as appropriate, either to the most current version (IMSO Convention (2008)) or to the version that took effect as of the time of privatisation (IMSO Convention (1999)).

3 The second agreement was an agreement known as the Operating Agreement, which was signed by the designated national operators or Signatories that provided the actual financing for the organisations. Commercial oversight was provided by the Signatories through management boards which functioned more like partnership committees than corporate boards of directors. These were known respectively as the INTELSAT Board of Governors, the INMARSAT Council, and the EUTELSAT Board of Signatories. INTELSAT had one additional organ, known as the Meeting of Signatories, which met annually and was similar to a corporate shareholders meeting. The commercial governance structure as

process, which provided for a degree of consistency notwithstanding some of the key differences among the three organisations.

For example, while both INTELSAT and INMARSAT were globally-based treaty organisations, the extent of developing world participation in the INTELSAT privatisation process far exceeded that which was present in the INMARSAT privatisation process. The number of developing world countries in INMARSAT was less than twenty, whereas in INTELSAT the number of developing countries that were members was more than double that number. In the case of EUTELSAT, as contrasted with both INTELSAT and INMARSAT, the range of governmental participants was significantly narrower, excluding a number of major players outside of Europe that were actively involved in the INTELSAT or INMARSAT privatisation debates, such as the United States, Canada, Mexico, Brazil, Russia, Japan, Malaysia, Australia, Kenya, Cameroon and South Africa, just to name a few. Moreover, not unexpectedly, developing world involvement in the EUTELSAT privatisation process was essentially non-existent as well.

The privatisation deliberations within both INTELSAT and INMARSAT began in the early 1990s in a somewhat rudimentary form, while meaningful consideration of the possible full privatisation of either organisation did not seriously surface until later in the decade. This then coincided with the same general timeframe in which the privatisation of EUTELSAT was considered. The full privatisation of INMARSAT was completed on April 15, 1999; the full privatisation of EUTELSAT (or completion of the "Transformation" which was the operative terminology used in the EUTELSAT context) was completed on July 2, 2001; and the full privatisation of INTELSAT was completed on July 18, 2001.[4]

For INMARSAT, the initial consideration of privatisation was driven by the desire to enter the emerging hand-held mobile satellite telephony market, following in the footsteps of the ambitious business plans advanced by Iridium, Globalstar and certain other companies. As events unfolded,

---

an embedded part of the ISOs fell away at the time of their privatisation, with the formal termination of the Operating Agreements and the transformation of the investment shares held by Signatories into shares of stock in the privatised companies.

4 For each of the ISOs, at the time of their respective privatisations, the main intergovernmental treaty agreements remained in force but were the subject of significant amendments, with those amendments taking provisional effect as of the effective date of privatisation (corresponding as well to the date on which each organisation's Operating Agreement was terminated). The most current version of each of the intergovernmental treaty agreements can be found on the respective internet home pages for each organisation, www.itso.int, www.imso.int, or www.eutelsatigo.int, respectively.

it proved politically impossible for INMARSAT to enter this market directly in its capacity as an international organisation, but instead its pursuit of this business opportunity was redirected towards the creation of an entirely separate company to be known as ICO Ltd. (ICO), partially owned by INMARSAT in its own right as well as by a number of INMARSAT Signatories, which came to fruition in 1995. The deliberative process leading up to ICO's formation unleashed broader issues about INMARSAT's future role, which was subsequently transformed later that year into a more focused pursuit of privatisation.

For INTELSAT, serious consideration of the possibility of a major restructuring first surfaced in 1993, but was quickly diverted from consideration of full privatisation into the possibility of creating a spin-off company, to which some of INTELSAT's more risky business activities could be transferred, along with a designated number of INTELSAT satellites on which such services could be provided. This was conceived primarily as a means of reducing some of the perceived longer term commercial risks facing INTELSAT's main mission of assuring global connectivity and coverage on a non-discriminatory basis. Additionally, it effectively functioned as a test case for the consideration of more comprehensive restructuring that would occur later. This initial effort culminated in 1998 with the creation of a spin-off company from INTELSAT, which came to be known as New Skies Satellites.[5] However, rather than quelling the impetus for restructuring the remainder of INTELSAT, this development served as the catalyst for the subsequent consideration of the full privatisation of what remained of INTELSAT which began in earnest later that year.

While EUTELSAT did not undergo a similar two-step process, the political, regulatory and commercial pressures that had ultimately led both INMARSAT and INTELSAT in the direction of full privatisation were definitely being felt within the EUTELSAT context as well. EUTELSAT's restructuring was also heavily influenced by the European Commission's increasingly active participation in the INTELSAT and INMARSAT privatisation efforts. Consequently, EUTELSAT undertook more serious consideration of full privatisation in roughly the same timeframe as INTELSAT, with an initial decision in principle to privatise taken at a meeting of the EUTELSAT Assembly of Parties in May 1999.[6]

---

5 INTELSAT Assembly of Parties, *Record of Decisions of the Twenty-Second (Extraordinary) Meeting*, AP-21-3E (30–31 March 1998).

6 EUTELSAT IGO, *EUTELSAT IGO Amended Convention*, Editorial Note.

Although deliberations conducted within each of the three organisations were separate and distinct, the sequencing in which the three ISOs conducted their internal debates, which overlapped to some extent, had a definite impact on their ultimate outcome. Because the full privatisation of INMARSAT was completed prior to the full privatisation of INTELSAT or EUTELSAT, in many respects the INMARSAT privatisation process functioned as a roadmap of sorts for what was later addressed within INTELSAT and EUTELSAT, including the critical issue of whether it would be necessary to retain any form of intergovernmental organisation post-privatisation. At the same time, the fundamental character of the underlying public service obligation – which the preservation of some form of international oversight was intended to serve – turned out to be substantially different in the case of each of the three privatised entities.

### 1.2. *Preconditions to Privatisation*

While the impetus for privatisation was driven by a mixture of political and economic concerns, in each instance the question of whether and how the underlying public services mission that each of the organisations was created to serve would be maintained in the event of privatisation loomed as a precondition that had to be addressed as a prelude to any further consideration of privatisation.

In the case of INMARSAT, this precondition took the form of determining whether and how the privatised Inmarsat company would maintain Global Maritime Distress and Safety System (GMDSS) functionality following privatisation. This touched upon what was fundamentally a non-commercial aspect of Inmarsat's business operations and, as such, it was not directly related to the terms and conditions of the commercial services that the privatised Inmarsat would provide.

Similarly, in the case of EUTELSAT, the critical precondition centred on whether and how the privatised Eutelsat company would be required to maintain full pan-European coverage. While this concern may have touched upon the terms and conditions of the commercial services that the privatised Eutelsat company would provide, again it did not have a direct material effect on those terms and conditions.

It was only with respect to INTELSAT that the necessary preconditions would have a fairly direct and dramatic impact on the terms and conditions by which the privatised Intelsat company would provide commercial services, at least to a select segment of its customer base. In the case of

INTELSAT, this impact was unavoidable, given the nature of the public services mission that had been originally entrusted to INTELSAT when it was first established on an interim basis as an international organisation in the 1960s and which then carried through to when it was more formally established as a treaty-based organisation in 1973. In particular, the focal point of that mission was the provision of affordable and ubiquitous commercial services to those users that were primarily dependent on INTELSAT for global connectivity.

### 1.3. *The Fate of the Public Services Obligations*

The approach ultimately selected in each instance centred on a fundamental decision to transfer the purely commercial operations previously performed by the ISOs to newly established national law companies that would operate without benefit of any of the special protections otherwise accorded international organisations. As such, these national law companies would be totally lacking in any intergovernmental character whatsoever. With respect to policy formulation activities and public services obligations performed by each of the intergovernmental organisations, the decision was made to preserve these functions by retention of an international organisation structure, albeit on a scaled-down basis in terms of both its mission and size – in other words, by means of what could best be described as a residual intergovernmental international satellite organisation (RIISO).[7]

These RIISOs would not be newly created bodies, but would instead function as continuations of the original ISOs, with their commercial activities stripped out. Moreover, their mandates were also reformulated into a narrower and more focused mission related to assuring that certain public service obligations previously served by the ISOs would be properly maintained by the resulting privatised commercial entities. To effectuate this, the underlying treaty establishing each organisation was significantly amended to delete those provisions that related to commercial operations and to narrow and re-state the principal mission to be served by the organisation. These amended agreements took effect as of the date of privatisation, generally through a process known as provisional application,

7 The term RIISO will hereinafter be used specifically to refer to the international organisations post-privatisation, whereas the term ISO will continue to be used in the case of references that are made to the international organisations prior to privatisation.

a doctrine that allowed the amendments to become provisionally effective prior to the time that the treaty amendment formalities had been fully adhered to.[8]

In addition to substantially revised charters, the organisations themselves assumed new identities, at least in terms of how they were publicly identified. Thus, in the case of INMARSAT, following privatisation the RIISO that remained would continue to be known as the International Mobile Satellite Organisation, but re-labelled using a new acronym of IMSO. Similarly, in the case of INTELSAT, following privatisation the RIISO that remained would continue to be known as the International Telecommunications Satellite Organisation, but re-labelled as ITSO. Lastly, in the case of EUTELSAT, following privatisation the RIISO that remained would continue to be known as the European Telecommunications Satellite Organisation, but re-labelled as EUTELSAT IGO.

## 2. *RIISOs in a Post-Privatisation Environment: The Underlying Rationale for Preservation of International Oversight*

As each of the three major ISOs contemplated privatisation in the 1990s, the principal policy debate focused on how the privatisation process would affect the various public service obligations that were imbedded in their respective charters and were at the core of their *raison d'être*. The key question posed was whether, once privatisation occurred, there would be any continuing need to have an intergovernmental apparatus involved at all, or if that could simply disappear as part of the overall privatisation process. In other words, the overriding concern was whether and how the public service obligations that had previously been served by the ISOs should be maintained following privatisation, especially if one of the main purposes of the privatisation process was to fundamentally alter the intergovernmental character of the ISOs in terms of their governance and mission.

Not surprisingly, a fairly broad range of views on this critical issue was expressed, representing options that each carried with it a unique set of

[8] See David Sagar, "Provisional Application in an International Organisation," *Journal of Space Law* 27, no. 2 (1999): 99–116. Available: http://www.spacelaw.olemiss.edu/JSL/Back_issues/JSL%2027-2.pdf.

advantages and disadvantages. At one end of the spectrum, there was the option of doing nothing. This option was predicated on the neo-liberalist belief that market forces alone would be more than sufficient to safeguard these interests that had previously been protected by the ISO structure (that is, provision of affordable and ubiquitous commercial services to those users that were primarily dependent on INTELSAT for global connectivity), to the extent that such protection would even be necessary following privatisation. This view was fully consistent with and a natural outgrowth of the major impetus pushing for privatisation of the ISOs in the first place, which was the belief that having intergovernmental organisations as players in the international satellite telecommunications marketplace, particularly once meaningful competition had emerged among purely private sector players, was anathema to the proper functioning of a competitive marketplace and was blatantly anticompetitive. This view further held that, even a scaled-down ISO with no real operational role, could still potentially distort marketplace forces in ways that could adversely affect competition.

This certainly was the position of the US Party and possibly other like-minded countries that were troubled by the competitive imbalance potentially posed by the ISOs. However, irrespective of the merits (or lack thereof) of any such view, as a practical matter it ultimately proved completely untenable because of the impossibility of crafting the necessary, broad based consensus among the member States in any of the three ISOs to agree to privatisation if the underlying public service obligations were completely forsaken.

A second approach that was considered involved the formulation of a set of specific protective mechanisms that would allow for continued recognition and protection of these public service benefits, but which would not necessarily require the continuing existence of an international organisation in order to be effective. Instead, this approach would rely on the development and application of enforceable legal obligations imposed by means of contractual relationships, with enforcement mechanisms and remedies provided, in the case of alleged breaches, by a pre-agreed adjudicative process. This approach was predicated on the belief that any obligation imposed on the privatised entity to continue to fulfil certain pre-agreed public service obligations could effectively be enforced by some form of contractual arrangement that would be entered into as part of the privatisation process in the absence of any continuing intergovernmental supervision or oversight. In particular, this approach was initially advocated by the INTELSAT Board of Governors and the INTELSAT Meeting of Signatories, particularly in the very early days in which full privatisation of INTELSAT was under consideration.[9]

This approach certainly had a somewhat broader appeal than the approach advocating no oversight or supervision at all, particularly to the extent of its formal recognition of the need to in some fashion preserve the public service obligations embedded in the prior treaty arrangements, even in the absence of an intergovernmental organisation. However, a large number of INTELSAT Parties, including those representing countries most dependent on INTELSAT's services, remained fearful that the absence of any form of intergovernmental supervision over the privatised entity left too great a chance that abuses would go un-remedied. Thus, while this approach was not totally discarded, it was never seriously viewed as completely satisfactory by itself to afford the full range of protections warranted.

The third approach, and the one that was ultimately adopted, involved the formal preservation of an intergovernmental apparatus in the form of a RIISO, with specifically delineated oversight responsibilities, albeit with a scaled down mission as compared to that previously conferred on the ISOs prior to privatisation.

As noted above, variations of these options were considered by each of the three ISOs undergoing privatisation, but ultimately each separately arrived at the same conclusion that preservation of some form of intergovernmental apparatus would be desirable. As a practical matter, however, once INMARSAT, which was the first to squarely address this issue, opted for preservation of a RIISO, that decision effectively cast the die for the other two organisations as well. Indeed, it would have been extraordinarily difficult for either INTELSAT or EUTELSAT to eschew such an approach, once the Parties to the INMARSAT Agreement had accepted this approach as reasonable and appropriate.

Thus, having made the fundamental determination to preserve the international organisation apparatus, the stage was then set to address a number of key issues relating to the exact role and function that the surviving RIISOs would then play.

---

9 INTELSAT Assembly of Parties, *Report of the Board of Governors to the INTELSAT Assembly of Parties on INTELSAT Restructuring*, AP-24-14E, at para 8(b)(ii) (16 September 1999): 8.

## 3. *Essential Characteristics and Functions of the RIISOs*

### 3.1. *Core Principles*

Having decided to retain some form of intergovernmental organisation, the next issue concerned articulation of the specific purpose or purposes that the particular RIISO would serve, or – in other words – its fundamental mission. In the broadest sense, the mission of each RIISO encapsulated the underlying policy mandate of the respective ISO from which it had evolved.

In the case of ITSO, the main purpose, as set forth in Article III of the ITSO Agreement, was to ensure that the privatised Intelsat would provide, on a commercial basis, international public telecommunications services in order to ensure the performance of the following three Core Principles: (i) to maintain global connectivity and global coverage; (ii) to serve lifeline connectivity customers, and (iii) to provide non-discriminatory access to the Company's system.[10] The Core Principles had further been embedded in the initial By-laws of Intelsat Ltd. at the time of its creation. Additionally, pursuant to Article V of the ITSO Agreement, ITSO was further charged with the obligation to take all appropriate actions in order to supervise the performance by the privatised Intelsat in relation to the Core Principles.[11] Particular emphasis was placed on safeguarding the principle of non-discriminatory access to the privatised Intelsat's satellite system for existing and future public telecommunications services offered by it when space segment capacity was available on a commercial basis. Intelsat's adherence to these Core Principles was more generally referred to as the discharge of its public service obligations.

The Primary Purposes for IMSO, as set forth in Article 3 of the IMSO Convention, were to ensure that the privatised Inmarsat company adhered to the following basic principles: (i) to ensure the continued provision of GMDSS satellite communications services, in particular those that are specified in the International Convention for the Safety of Life at Sea (SOLAS) and the Radio Regulations of the International Telecommunication Union (ITU); (ii) to provide services without discrimination on the basis of

---

10 Art. III, ITSO Agreement

11 Art. V, ITSO Agreement.

nationality; (iii) to act exclusively for peaceful purposes; (iv) to seek to serve all areas where there is a need for mobile satellite communications, giving due consideration to the rural and the remote areas of developing countries; and (v) to operate in a manner consistent with fair competition, subject to applicable laws and regulations.[12]

However, following further amendment of the IMSO Convention in 2008, the first purpose was modified to delete references to the SOLAS Convention and the ITU Radio Regulations, substituting instead that this was to be done according to the legal framework set up by the International Maritime Organisation (IMO). Moreover, the second, fourth and fifth principles were deleted in their entirety, substituted with the more general purpose of performing the oversight functions in a fair and consistent manner among Providers.[13] In addition those amendments added a new Article 4 entitled "Other Functions", which authorises IMSO, subject to further decision of the Assembly of Parties, to assume functions and duties as Coordinator for the Long Range Identification and Tracking (LRIT) System, a maritime tracking system that had been established by the IMO. This coordination function would be undertaken in accordance with decisions of the IMO, and would be carried out by IMSO in a fair and consistent manner, at no cost to IMSO Parties and subject to their express authorisation to do so.[14]

The primary purpose for EUTELSAT IGO, as set forth in Article III of the EUTELSAT Amended Convention, was to ensure the observance of certain Basic Principles by the privatised Eutelsat. These Basic Principles included the following: (i) meeting its public service/universal service obligations, applicable to the space segment and its use for the provision of services connected to the public switched telephone network, with audio-visual services and future services to be provided in conformity with relevant national regulations and international agreements, including the provisions of the European Convention on Transfrontier Television; (ii) provision of

[12] Art. 3, IMSO Convention. The language concerning use of the system for peaceful purposes was also contained in the original INMARSAT Convention adopted in 1979 (Art. 3(3), INMARSAT Convention), although historically that language has been accorded a fairly liberal interpretation and in particular was never seen as an impediment to possible use of INMARSAT space segment capacity by military or national defense customers in member countries.

[13] Art. 3, IMSO Convention (2008).

[14] Art. 4, IMSO Convention (2008). *See infra* at Section 4.3.

pan-European coverage, pursuant to which the privatised Eutelsat shall, on an economic basis, seek to serve all areas where Member States have a need for communications services; (iii) maintaining non-discrimination, with services to be provided to users on an equitable basis subject to commercial flexibility and consistent with applicable laws; and (iv) fair competition, meaning that the privatised Eutelsat shall comply with all applicable laws and regulations relating to fair competition.[15] Additionally, the EUTELSAT IGO was charged with ensuring continuity under international law (and in particular the ITU Radio Regulations) of the rights and obligations associated with the use of frequencies transferred to the privatised Eutelsat that were derived from its subsequent operation of the EUTELSAT space segment.[16]

### 3.2. *Scope of Authority and Powers*

In order for the RIISOs to fulfil their basic missions, these organisations had to be accorded certain powers and authority to act. The determination of what powers and authority would be accorded ultimately required the application of a fairly delicate balancing exercise, tempered by the disparate views as to the necessity for the organisations to even exist. The primary issue to be addressed was whether the RIISOs should be accorded any degree of independent supervisory oversight over the privatised entity, or whether the nature and scope of their authority would be primarily – if not exclusively – dependent upon particular written protocols in which the exact nature of the relationship between the RIISO and the privatised entity would be specifically defined. In all three instances, the latter approach was taken.

Generally, for all three organisations, the primary source of authority came from a specific written agreement entered into between the RIISO and the respective privatised entity. In the case of both ITSO and IMSO this document was referred to as the Public Services Agreement (PSA); in the case of EUTELSAT IGO it was referred to as the Arrangement.

For ITSO, the real authority that it had over the privatised Intelsat arose almost entirely from the terms of the ITSO PSA. While the ITSO Agreement characterised ITSO's role as "supervisory" in nature, ITSO was not accorded

15 Art. III(a), EUTELSAT IGO Amended Convention.
16 Art. III(b), EUTELSAT IGO Amended Convention.

any independent supervisory powers over Intelsat other than what was contained in the ITSO PSA. Rather, ITSO's supervisory role was characterised as its entitlement to review and assess Intelsat's performance in relation to its public service obligations.[17]

Even with respect to the special relationship that had been established between Intelsat and the user community that was most dependent upon its satellite services in order to meet their most basic international public telecommunications services, which was referred to as the Lifeline Connectivity Obligation (LCO), this relationship was also fundamentally contractual in nature. The terms and conditions establishing the contours of the relationship between Intelsat and any LCO customer (LCO Customer) were contained in a separate document known as the LCO contract (LCO Contract), and ITSO's role therein is largely advisory. At the request of a LCO Customer, ITSO may assist in the resolution of disputes related to an LCO Contract by advising the LCO Customer as to possible methods of resolving that dispute; the desirability of seeking arbitration under the LCO Contract; and on preliminary matters preparatory to arbitration, including such matters as the selection of experts and arbitrators. ITSO may also assist by conciliating disputes between Intelsat and an LCO Customer. ITSO, however, was specifically precluded from filing or in any way making arbitration claims on behalf of any LCO Customer or acting as an arbitrator in any dispute.[18]

In the case of IMSO, a formal PSA also served as the principal grounding for the relationship between IMSO and the privatised Inmarsat, similar in many respects to the PSA entered into between Intelsat and ITSO.[19] There is, however, one very important additional tool accorded to IMSO that was not accorded to ITSO, which is that of a "special share" in the privatised Inmarsat, which is incorporated into the Articles of Association of Inmarsat Holdings Ltd. This special share, which IMSO can hold for so

---

17 INTELSAT Assembly of Parties, *Record of Decisions of the Twenty-Fifth (Extraordinary) Meeting*, AP-25-3 (27 November 27 2000) (Attachment No. 3, Public Services Agreement Between Intelsat, Ltd., Intelsat Service Corp. and Art. 3, ITSO Agreement) (hereinafter referred to as ITSO PSA).

18 Ibid.

19 Public Services Agreement Between the International Mobile Satellite Organisation, Inmarsat One Ltd and Inmarsat Two Company (April 1999) (hereinafter referred to as IMSO PSA). Available: http://www.imso.org/pdfs/Public/Basic%20Documents/Public%20Services%20Agreement/P%20-%20Public%20Services%20Agreement%20-%20English.pdf.

long as the IMSO PSA is in existence, unless it voluntarily decides to redeem it prior thereto, accords IMSO as the holder of the special share, the absolute veto power over any effort by the privatised Inmarsat to modify its Memorandum of Association by removing or altering language pertaining to its obligation to adhere to the Basic Principles or to undertake the voluntary winding up of the company for any reason other than insolvency. The special shareholder is entitled to attend any general meeting of shareholders and to speak at such meetings, but shall not have the right to vote other than with respect to the specific authority conferred by the terms of the special share.[20] While the actual scope of authority conferred by this special share is somewhat limited, it nonetheless represents as least the ability of IMSO to directly intercede in the affairs of the privatised Inmarsat.

Perhaps the most interesting aspect of the special share arrangement was that it did not serve as precedent for the subsequent INTELSAT privatisation. In part, the reason why this approach proved palatable in the case of IMSO but not ITSO was that the concept of a special share held by a regulator (in most cases the government following a telecom privatisation) was fairly solidly imbued in the European legal landscape but not in the US landscape, where inclusion of such a mechanism in the case of the INTELSAT's privatisation would have been extremely problematic. The fact that the INMARSAT privatisation had effectively been completed prior to passage of the legislation in the United States intended to help shape the manner in which the ISOs were to be privatised, known as the Open-Market Reorganisation for the Betterment of International Telecommunications Act (ORBIT Act)[21] in 2000, also afforded IMSO a greater degree of legitimacy with respect to possession of a special share than could ever have been accorded to ITSO in a similar context. Thus, while the type of special share held by IMSO was clearly repugnant to the philosophy underlying the ORBIT Act, as a practical matter it would have been impossible to retroactively challenge it in the ORBIT Act.

The situation was similar as well in the case of EUTELSAT IGO, where the primary supervisory authority emanated from the document known as the Arrangement. The one critical difference between the Arrangement

20 Inmarsat, *Articles of Association of Inmarsat Holdings Ltd.*, Company No. 3674573, Incorporated 20 November 1998, as adopted by special resolution passed on April 8, 1999.

21 Open-Market Reorganisation for the Betterment of International Telecommunications Act, 47 U.S.C. § 761 (2000).

and the two PSAs is that the Arrangement has been amended on multiple occasions following EUTELSAT's privatisation, where this has not occurred in either of the other two organisations. As a result of these modifications, the EUTELSAT IGO appears to be the only RIISO that has in fact been able to expand the EUTELSAT IGO's supervisory responsibilities over time and to secure improvements to its funding. This was made possible by concerns expressed regarding the phased buyout of the privatised Eutelsat by various private equity groups and the potential impact this might have on the continued commitment of the privatised Eutelsat to adhere to the principal objectives or to preserve the essential European character of the organisation.[22]

### 3.3. *Continuity/Duration*

Once the basic decision to retain an RIISO had been reached, the duration or anticipated lifetime of the RIISO represented one of the most difficult and contentious issues that arose. Should the RIISOs be permitted to continue in perpetuity or should there either be a timeframe or a set of circumstances in which their existence could be terminated? The resolution of this issue was integrally related to the separate issue of funding.

In the case of IMSO, while no specific lifetime is specified in the amended Convention, a means by which the IMSO PSA can be terminated is provided under a variety of circumstances. These include: (i) by written agreement between IMSO and the privatised Inmarsat at any time; (ii) by written notice given by IMSO to the privatised Inmarsat at any time; or (iii) by written notice given to IMSO by the privatised Inmarsat indicating that the SOLAS Convention had been amended so as to provide that the GMDSS requirements contained therein could be satisfied by the carriage of ship earth stations operable with one or more other global satellite systems. In this latter case, termination will be effective three years after the notice is given or when the referenced amendment formally enters into force or when the IMO determines that the GMDSS requirements are being satisfied by other satellite system operators, whichever occurs later. The effective date of termination may be earlier if otherwise agreed by IMSO and the privatised Inmarsat.[23]

---

[22] *See infra*, section 4.2.

[23] Art. 18, IMSO PSA.

While the termination of the IMSO PSA does not automatically terminate the IMSO Agreement, it would terminate the privatised Inmarsat's obligation to fund IMSO, which would seem tantamount to effective termination of the IMSO Agreement. Of course, this approach did not anticipate the possible emergence of alternate funding mechanisms, which in the case of IMSO has now occurred in light of the recent expansion of IMSO's functional responsibilities to perform LRIT obligations, and which functions would not be funded by the means used to fund IMSO's public service obligations in any event.[24] As such, IMSO may have achieved some degree of potential immortality which may not necessarily be available to either of the other two RIISOs.

In the case of ITSO, the issue of duration took a number of interesting twists and turns. Very early on in the INTELSAT privatisation process, contemplation was given to the possibility that ITSO might only need to exist for a relatively short transitional period, perhaps as short as four years. However, given that the LCO Contract had a defined duration of twelve years, this view soon gave way to the notion that ITSO's lifetime should in some fashion be tied to the lifetime of the LCO Contract. Moreover, this scenario then lent support to the proposition that, once the LCO Contracts were set to expire, there might not be any compelling rationale for ITSO to continue to exist, thereby supporting the notion that ITSO's duration should be coterminous with, but not longer than, the duration of the LCO Contracts.

This was certainly the view of the governments that were most resistant to the concept of an RIISO in the first place. However, within the developing world, the possibility of a fixed duration for ITSO was viewed with considerable alarm, coupled with the belief that ITSO could continue to be of value even after the LCO Contracts had come to an end. Furthermore, although the LCO arrangement would come to an end after twelve years (as they have in July 2013), individual LCO commitments made prior to the expiration of the program could continue in force far beyond the expiration date. This provided a further justification for the continued existence of ITSO beyond the projected expiration date for the LCO arrangement. At a very minimum, these Parties believed it would be premature and inadvisable to make definitive judgments regarding ITSO's lifetime so far in advance of the termination of the LCO arrangement.

---

24 *See infra*, section 3.3.

As such, a strong preference emerged among the developing world Parties to defer the final determination regarding the duration of ITSO's ongoing existence until at least the nominal end date for the LCO Contracts. This resulted in a conceptual approach that prevailed throughout most of the time period in which the particulars of the Intelsat privatisation were being debated. Under this conceptual approach, a minimum duration time period of 12 years for ITSO's duration was firmly set, while allowing for the possible extension of its lifetime beyond that time period, should the Assembly of Parties take a formal decision at the end of the twelve year period that the organisation should continue.[25]

The practical effect of this approach was that an affirmative decision by a two-third's majority of the Parties to continue ITSO's lifetime beyond the twelve-year period would be required for its continuation. While this avoided setting a firm date by which ITSO would cease to exist, this approach was widely seen as creating a fairly substantial hurdle to the likely continuation of ITSO beyond the expiration of the twelve year period, given the potential difficulty of mustering a two-third's majority vote in favour of its continuation.

However, at the October 2000 INTELSAT Assembly of Parties, when the entire privatisation package came up for final consideration, the developing world Parties, taking full advantage of the considerable leverage they then held at that point, launched a final major assault on this issue. Their objective was to reverse the basic presumption as to how the decision whether to continue ITSO's existence would be made, such that it would require a two-thirds majority decision to terminate ITSO at that time. In the absence of such a two-thirds majority decision, ITSO would continue to exist until such future time as a decision to terminate it was made, again requiring a two-thirds majority vote to approve such termination. This effort proved successful, with the practical effect of significantly increasing the difficulty of terminating ITSO at the end of twelve years and thus enhancing the likelihood that ITSO would continue in existence beyond such a time.[26]

---

25 INTELSAT Assembly of Parties, *Report of the Penang Working Party to the Twenty-Fifth (Extraordinary) Assembly of Parties*, AP-25-7E, at para. 13(a)(viii) (27 June 2000): 4.

26 INTELSAT Assembly of Parties, *Record of Decisions of the Twenty-Fifth (Extraordinary) Meeting*, AP-25-3E, at para. 8(a)(vii) (27 November 2000): 7; *see* INTELSAT Assembly of

The issue of whether ITSO would continue to exist beyond the twelve year period for which it was initially established has now been squarely addressed and resolved at a meeting of the ITSO Assembly of Parties held in Kampala, Uganda in July 2012. At that meeting, the decision was taken to continue the organisation until at least July 18, 2021, and that further consideration of possible termination of the ITSO Agreement would be taken up at the ordinary meeting of the ITSO Assembly of Parties that would take place in 2020.[27] As before, any decision to terminate the ITSO Agreement would require a two-thirds majority vote of those Parties in attendance at an Assembly of Parties meeting.

In reaching the decision not to terminate the Agreement in 2013 (which would have been the earliest permissible termination time consistent with Article XXI of the ITSO Agreement),[28] the Assembly of Parties expressly supported the following three conclusions that had been presented in a Report of the Director General on the Future of the ITSO Agreement:[29]

- That the performance by Intelsat of its Public Service Obligations remains at least as important today as it was in 2001;
- That the ITSO Agreement plays a crucial role in ensuring that the principles, obligations and objectives as defined during the restructuring continue to be adhered to; and
- That it is essential to continue to have a mechanism in place for upholding and protecting the Public Service Obligations as well as the Parties Common Heritage Assets.[30]

---

Parties, *Summary Minutes of Discussions of the Twenty-Fifth (Extraordinary) Meeting*, AP-25-4, at paras. 77–89 (13–17 November 2000): 16–17 (remarks of the representatives of the Parties of France, Gabon, China, Mexico, Cameroon, Togo and Morocco).

27 ITSO Assembly of Parties, *Record of Decisions of the Thirty-Fifth Meeting*, AP-35-3E, at para. 13 (3–6 July 2012).

28 Art. XXI (Duration), ITSO Agreement, provides that: "This Agreement shall be in effect for at least twelve years from the date of the transfer of ITSO's space system to the Company. The Assembly of Parties may terminate the Agreement effective upon the twelfth anniversary of the date of transfer of ITSO's space system to the Company by a vote pursuant to Article IX(f) of the [Agreement]. Such decision shall be deemed to be a matter of substance."

29 ITSO Assembly of Parties, *Report of the Director General on the Future of the ITSO Agreement*, AP-35-7E (31 May 2012).

30 ITSO Assembly of Parties, *Record of Decisions of the Thirty-Fifth Meeting*, AP-35-3E, at para. 12 (3–6 July 2012).

In the case of EUTELSAT IGO, there again is no specific timeframe identified in either the EUTELSAT IGO Amended Agreement or the Arrangement. Article XIV of the Amended Agreement does provide that the Parties may bring EUTELSAT IGO to an end by termination of the Convention by a two-thirds majority of all Parties. However, no decision to terminate EUTELSAT IGO shall be taken as long as the international rights and obligations for use of frequencies derived from the operation of the EUTELSAT space segment transferred to the privatised Eutelsat have not been entirely extinguished, unless otherwise agreed to by the privatised Eutelsat.[31] As a practical matter (and presumably reflecting the intent of the Parties), this is quite unlikely to occur, because terminating EUTELSAT IGO would result in the cancellation of all orbital rights in existence at the time of privatisation.

### 3.4. *Funding Mechanism(s)*

An issue that is inextricably linked to duration is funding, because without a funding mechanism in place, duration is fundamentally a moot point. One way or another, funding for the RIISOs ultimately had to be borne by the privatised companies, because it was clear that the governments had no interest whatsoever in providing funding themselves. Exactly how that was to be levied on the privatised company, however, was subject to somewhat different approaches being taken, where the approach taken in Inmarsat and Eutelsat differed fairly substantially from that taken in Intelsat.

For both IMSO and EUTELSAT IGO, the funding was set at a specified level, with the obligation imposed on the privatised company to provide the funding on an annual basis. In the case of Inmarsat, this was set by the IMSO PSA, which obligated the privatised entity to pay to IMSO the sum of £ 300,000 (roughly US$ 450,000 annually based on then-prevailing exchange rates) for operational expenses, with the further proviso that any funds unexpended or uncommitted were to be credited against the payment to be made by the Company for the following year. Moreover, the privatised Inmarsat was obligated to enter into good faith negotiations with IMSO over additional payments should IMSO face any increased annual costs or any unexpected costs arise. Additionally, a contingency fund of £ 100,000 (roughly US$ 150,000) would be set aside to meet IMSO's cost of arbitration or legal proceedings in connection with enforcement proceedings taken by

[31] Art. V, EUTELSAT IGO Amended Convention.

IMSO under the IMSO PSA, with a consultation mechanism also included for possible replenishment of the contingency fund in the event that a substantial part of the fund was utilized by IMSO.[32]

The funding levels for the EUTELSAT IGO were not generally publicly disclosed, although it does appear as though the EUTELSAT IGO Arrangement was subsequently amended to increase the funding levels, when the original funding level established proved insufficient for the proper discharge of EUTELSAT IGO's mandate. The one year for which the funding level has been publicly disclosed was 2005, for which the funding level was approximately € 700,000 (roughly US$ 840,000 at then-prevailing exchange rates).[33]

Setting the annual funding level at a fixed amount was seen as a means of preventing uncontrolled growth of the RIISO, particularly in terms of staffing. Also, since the duration of the organisation was somewhat indefinite in nature, providing for funding on a year-to-year basis seemed to represent the most sustainable approach.

In the case of ITSO, the dynamic played out in a somewhat different fashion. By the end of the privatisation process, the likelihood that ITSO might continue to exist for longer than twelve years had significantly increased. However, throughout much of the privatisation process, it had seemed more likely that ITSO's duration would be only for a twelve-year period, hence a funding approach tied to providing assured funding for a twelve-year period seemed appropriate and logical. As such, the decision was made to arrange for the funding for the full twelve-year period in advance, which would be handled by an annuity funded by Intelsat at the time of privatisation. The initial amount of funding for the first year was set at US$ 1.2 million, with an annual adjustment, largely viewed as covering inflation, of 3 per cent per year. This meant that, by year twelve, the funding amount for that year would be approximately US$ 1.8 million, with the year-to-year funding levels for all intervening years determined based on a linear progression between these two figures.[34]

In addition to this funding, it was recognised that in the event that ITSO found itself in a major dispute with Intelsat for any reason, the normal

32 Art. 15, IMSO PSA.

33 Eutelsat, *Offering Memorandum*, Eutelsat Communications (29 November 2005): 66. Available: http://www.eutelsat.com/investors/pdf/regulated-info/IPO-Final-OM.pdf.

34 INTELSAT Assembly of Parties, *Record of Decisions of the Twenty-Fifth (Extraordinary) Meeting*, AP-25-3E, at para. 8(g) (27 November 2000): 9.

funding mechanism might not provide it with sufficient resources to be able to address this issue. As such, a separate contingency fund of US$ 500,000 was established, which was earmarked specifically for use in the case of disputes arising between ITSO and Intelsat.[35]

However, once the possibility that ITSO might exist for longer than twelve years became more likely, it was necessary to consider the issue of how ITSO would be funded after the annuity had expired. The mechanism chosen here was to insert into the ITSO PSA a provision that, should ITSO continue beyond twelve years, which is now the case, the funding obligation would fall upon Intelsat, with the specific funding level to be determined based upon good faith consultations between Intelsat and ITSO. Exactly how this consultation would occur was left fairly open ended, other than the proviso that under no circumstances would Intelsat be required to fund on an annual basis at any dollar amount higher than had been the funding level at the end of year twelve. Also, if at the end of year twelve, should ITSO continue beyond that time, Intelsat would be obligated to replenish the contingency fund to its original level, as adjusted for inflation, should any of the funds in the contingency fund been expended prior thereto.[36]

### 3.5. *Orbital Resources*

Another difficult issue in the privatisation process concerned what would happen to the orbital resources that were held by the ISOs prior to privatisation. This unfolded against the backdrop of historic concerns that the ISOs had derived unfair competitive advantages arising from their favourable ability to secure orbital slots from the ITU.[37] Thus, in the interests of

---

35 Ibid.

36 Art. 14, ITSO PSA.

37 While the ISOs could not directly file for orbital locations with the ITU, the practice that had been adopted was for a designated country to serve as the nominal Notifying Administration, although its role was fundamentally that of a pass-through mailbox on behalf of the ISO, rather than truly functioning as a Notifying Administration as that term was more conventionally used. As such, the actual responsibility for deciding when and which orbital resources would be requested resided entirely with the ISO, as did responsibility for any coordination activities required under the ITU Radio Regulations. The role of the Notifying Administration in this regard was entirely passive in nature. In all three instances, the country that had been selected as the Notifying Administration was the same country in which the ISO's official headquarters had been located. Thus, in the case of INTELSAT, the United States served as the nominal Notifying Administration, for INMARSAT, it was the United Kingdom, and for EUTELSAT, it was France.

establishing a more level playing field, there was a strong impetus to remove any special treatment to be accorded the orbital slots that would be utilized by the privatised companies before the ITU.

To achieve this result, it would be necessary to have the orbital resources transferred to one (or more) national administration(s) that would serve as a Notifying Administration in exactly the same manner as occurred in the case of private satellite operators. Moreover, whatever happened with respect to existing orbital resources, it was clearly recognised that, in the case of any orbital resources acquired by any privatised entity following privatisation, it would be absolutely essential that those orbital resources be obtained through the normal actions of a Notifying Administration acting on behalf of the privatised company and the RIISO could have absolutely no involvement whatsoever in that process. Finally, the outstanding INTELSAT precedent that occurred with respect to the previous establishment of New Skies as a spin-off company from INTELSAT further supported this outcome. There, the Netherlands had been designated as the Notifying Administration for New Skies and, once that occurred, the Netherlands assumed full responsibility for all of the orbital resources that were transferred from INTELSAT to New Skies.[38]

In the case of the INMARSAT privatisation process, this proved to be a fairly simple decision, with the United Kingdom selected as the Notifying Administration to which the orbital resources were to be transferred.[39]

In the case of the INTELSAT privatisation process, while the final decision was to transfer the existing C and KU-band orbital resources to the United States and the orbital resources in all other frequency bands (KA, V-band) to the United Kingdom, these decisions proved quite controversial along the way, particularly the transfer of all existing C and Ku-band registrations to the United States.[40] Indeed, this was the only substantive issue that the body that had been established to develop the particulars of the privatisation plan (known as the Penang Working Party or PWP)[41] was

---

38 INTELSAT Assembly of Parties, *Record of Decisions of the Twenty-Second (Extraordinary) Meeting*, AP-21-3E (30–31 March 1998): 11. For purposes of the ITU Radio Regulations, this was treated as a transfer of responsibility from the nominal Notifying Administration (the United States) to the newly-designated Notifying Administration (the Netherlands).

39 INMARSAT Assembly of Parties, *Report of the Thirteenth (Extraordinary) Session of the INMARSAT Assembly*, Assembly/13/Report, at para. 4.1.13 (8 October 1998): 9.

40 INTELSAT Assembly of Parties, *Record of Decisions of the Twenty-Fifth (Extraordinary) Meeting*, AP-25-3E, at para. 29 (27 November 2000): 19.

41 The Penang Working Party was so named because it was created at AP-24, which was held in Penang, Malaysia.

unable to resolve satisfactorily in its work effort leading up to the final approval of the privatisation process.[42] As a result, the definitive arrangements with respect to the treatment of the orbital resources was not finally determined until the November 2000 INTELSAT Assembly of Parties meeting, and the corresponding Article in the revised ITSO Agreement was similarly not finally formulated until that time.

While, as noted above, the ultimate decision was to transfer INTELSAT's orbital registrations to the United States and United Kingdom, this was not done as cleanly as had been the case in INMARSAT or previously with the New Skies transfer of orbital registration to the Netherlands. First of all, the INTELSAT Assembly of Parties decided that the transfer could not occur until each selected Notifying Administration had ratified the amended treaty and had also issued licenses to the privatised company.[43] Moreover, the concept of the Parties' Common Heritage was formulated and inserted into the revised treaty, which was a concept that had not been utilized either in the case of the creation of New Skies or the INMARSAT's privatisation.[44] Finally, certain specific obligations were crafted that fell on any Notifying Administration selected that would directly affect how they were to conduct themselves as Notifying Administrations.[45]

The concept of the Parties' Common Heritage was predicated on the premise that the orbital resources and associated frequency spectrum assignments that were to be transferred to US and UK national administrations possessed a unique legacy that differentiated them from other orbital resources that were within the control of those Notifying Administrations, which warranted special attention and treatment. The significance of this was that, notwithstanding the transfer of the orbital locations to the United States and the United Kingdom, the Parties to the treaty agreement retained an ongoing interest in how those orbital locations were to be utilized and managed while otherwise ostensibly under the control of those Notifying Administrations. The inclusion of this concept in the revised treaty served as a critical safety net of sorts with respect to the future use and availability of those orbital resources to satisfy the Core Principles and, in the absence of such a measure it would be uncertain if the transfer of the orbital

---

42 INTELSAT Assembly of Parties, *Report of the Penang Working Party to the Twenty-Fifth (Extraordinary) Assembly of Parties*, AP-25-7E (27 June 2000): 50–52.

43 Art. XII(a), ITSO Agreement.

44 Artt. I(a)(l), XII(e), ITSO Agreement.

45 Ibid.

resources to the United States and the United Kingdom would have been approved.

As a result of this, the Notifying Administrations were given a unique set of obligations relating to their management of the orbital resources that differed in certain fundamental respects from the manner in which those Administrations would manage other orbital filings that they made with the ITU for other satellite operators, or for that matter in the case of any new satellite filings made on behalf of Intelsat following privatisation or which may have been incorporated as part of the Intelsat system by means of merger or acquisition. These obligations included the requirements that the Notifying Administrations had to: authorise the use of such frequency assignments by Intelsat so that the Core Principles may be fulfilled; report annually with regard to their adherence to their obligations; seek the views of the Director General regarding actions required to implement Intelsat's fulfillment of the Core Principles; work with the Director General to expand access to lifeline countries; notify and consult with the Director General on ITU satellite system coordinations that were undertaken on behalf of the Company to assure that global connectivity and services to lifeline users were maintained; and consult with the ITU regarding the satellite communications needs of lifeline users.[46]

The approach taken with EUTELSAT materially differed from both Inmarsat and Intelsat, in that the registration of the EUTELSAT orbital locations held as of the time of privatisation remained with the EUTELSAT IGO following privatisation.[47] Any additional orbital locations registered following privatisation would be registered by France as the Notifying Administration. The retention of the orbital resources within the EUTELSAT IGO would give it a degree of leverage over the privatised Eutelsat that greatly exceeded the leverage that either ITSO or IMSO would have over their respective privatised companies.

### 3.6. *The Special Case of Protection of Lifeline Connectivity Customers*

The most unique aspect of the ITSO Agreement that differentiates the role of ITSO from the other two RIISOs concerns the concept of LCO protection. Moreover, this concept, by providing a safety net for LCO Customers that

46 Art. XII(c), (e), ITSO Agreement.

47 Art. III(b), EUTELSAT IGO Amended Convention.

would otherwise be lost with privatisation, would serve as the critical lynchpin to securing the agreement of the developing world Parties to support privatisation of INTELSAT in the first place.

In seeking this form of protection, developing world Parties were motivated by three separate though interrelated concerns. Their first concern was that, following privatisation, the prices for the capacity those countries used at the time of privatisation could increase. This was driven by the fact that their satellite links represented the lighter density routes served by the INTELSAT system, which were not necessarily the most lucrative routes. Therefore, these countries feared that, based on purely economic considerations, Intelsat could well want to charge more to maintain such service following privatisation. In addition, since alternative sources of suitable satellite capacity were limited, the affected developing world Parties would have no choice but to accept the higher charges. To address this concern, some form of price protection would be necessary, which led to the formulation of the first element of the LCO protection – a firm guarantee that, for an extended time period (ultimately determined to be twelve years), prices for services being provided as of the time of privatisation to LCO Customers could not be increased.[48]

The second concern was that the specific satellite capacity these countries used at the time of privatisation could be diverted to other purposes, with inferior satellite capacity substituted in its place. One possible way that LCO Customers could protect themselves against this threat, even without any special protections, would be to place service orders for multi-year terms (which would have the secondary benefit of lower charges as well). However, in the case of many Parties falling within the category of LCO Customers, they either still operated as governmental departments functioning as postal, telegraph and telephone (PTT) authorities or possibly even as corporatized governmental business units. As such, these entities faced budgetary restrictions on them, which precluded them from ordering satellite capacity on a longer-term or multi-year basis, and in some instances restricted them to ordering capacity solely on a year-to-year basis. Such Parties wanted assurances that, as long as they continued to renew their capacity orders on a timely basis, the same capacity as they had previously received would continue to be available. This led to the

48 INTELSAT Assembly of Parties, *Record of Decisions of the Twenty-Fifth (Extraordinary) Meeting*, AP-25-3 (27 November 2000) (Attachment No. 4, LCO Contract, Section 4).

second element of the LCO concept, which was a guarantee that capacity would remain available as long as customers continued to order it. However, if for any reason, the LCO Customer cancelled the particular capacity order or failed to timely renew it, this protection would then be forfeited with respect to that specific capacity order (this would also be true in the case of the price protection as well).[49]

The third concern was that, while there would be protection against price increases, LCO Customers would be unable to benefit from any expected future price reductions, which had occurred with great regularity throughout INTELSAT's existence pre-privatisation. To address this concern, the decision was made to establish a basket of representative services (LCO Basket) that would serve as an indicative benchmark for pricing trends generally. If the commercial rates charged by the privatised Intelsat for those services falling within the LCO Basket decreased by more than 15 per cent over time, then all LCO-protected prices would be adjusted downward by 15 per cent as well. Moreover, once this had occurred, the pricing level associated with the LCO Basket would be reset, so if there were a subsequent further reduction of 15 per cent, then the LCO Customers would be entitled to a second such reduction. To monitor this, the privatised Intelsat was required to issue annual reports to ITSO concerning the pricing of items in the LCO Basket. It was further stipulated that the composition of the LCO Basket would be reviewed after the passage of four and eight years from the date of privatisation, to determine whether the representative services included therein continued to be appropriately representative of actual services that were being provided.[50]

With these elements in place, the basic contours of the LCO protection package came together. This then gave rise to the second issue, which concerned how to identify those countries or customers that should be eligible for LCO protection. Here the formulation used involved a combination of two considerations, the World Bank definition of low income (per capita GNP of less than US$ 755), and the ITU measure of low teledensity (less than three lines per 100 inhabitants). A total of 69 countries met one or both criteria and were therefore designated as eligible LCO Customers, which meant that all of their traffic carried on the INTELSAT system at the time of privatisation would be eligible for LCO

49 Ibid.

50 Ibid.

protection.[51] Recognizing that there could be specific circumstances which were not captured by these criteria, a petition process was also established by which any country or specific Intelsat customer that did not meet these criteria could still petition for eligibility for LCO protection, and those petitions were reviewed on a case-by-case basis. A total of 62 such petitions were filed, and of those 42 were granted (27 petitions were for all traffic carried by that country or customer on the INTELSAT system; 14 petitions were for specific links of a country or customer carried on the INTELSAT system, but not necessarily for all of the traffic for that country or user carried on the INTLSAT system; and 1 petition relating to special designation of LCO status on the U.N. for its use of INTELSAT's space segment). Altogether, a total of 111 countries or users were identified as eligible for LCO protection.[52] In addition, a specific provision was added that would allow any new country created after privatisation to apply for LCO protection as well as certain non-LCO countries which experienced an emergency-type situation to apply for LCO protection, even after privatisation.[53]

At the same time, participation in the LCO program was voluntary, and if a country chose not to avail itself of these protections, it would not be required to do so. Also, for any specific services that received LCO treatment, if following privatisation the basic parameters of those services were sufficiently modified that they no longer corresponded to the parameters applicable at the time of privatisation (that is, the amount of capacity covered by a lease was increased or the nature of the service offering utilized

[51] This designation was voluntary and, if for any reason a particular country did not want to avail itself of LCO protection on one or more links, it was free to have those links treated as simply regular commercial service provided by the privatised Intelsat which would be subject to Intelsat's standard commercial terms.

[52] INTELSAT Assembly of Parties, *Record of Decisions of the Twenty-Fifth (Extraordinary) Meeting*, AP-25-3E, at para. 9 (27 November 2000): 9–10; Attachment No. 5 (LCO Eligible Countries Based on World Bank Definition of "Low Income" (GNP/Capital less than US $765) and/or a Teledensity of Less Than Three as Defined by the ITU: Calendar Year 1999); Attachment No. 6 (Countries and Locations in Which Specific Commitments Are Recommended for LCO Protection, on the Basis of Petitions, Where the Country Has a GNP of US$5 Billion or Less); Attachment No. 7 (Countries and Locations in Which Specific Commitments Are Recommended for LCO Protection on the Basis of Petitions and There Are No Alternative Providers).

[53] To date, no such requests have been submitted.

was changed to a different service offering), then the various LCO protections would be lost.

While affording considerable protection, there were certain limits to the program. First of all, the protections applied only to those services that were actually being provided at the time of privatisation. All new services ordered after privatisation would have to be obtained on purely commercial terms. Second, the price and capacity guarantee protections were available only for services that were shorter than 10 years duration. For longer term services, because the pricing was already quite favourable, it was determined that no additional protection of this type would be necessary.

The final element was enforcement. The LCO arrangement was grounded in contract, with the expectation that, in the case of alleged violations by the privatised Intelsat, the remedy would be obtainable primarily through commercial measures. It was the LCO Customer that needed to invoke the dispute resolution mechanism with Intelsat directly, and as noted above, this process would unfold with limited if any direct involvement on the part of ITSO.[54]

### 3.7. *Other Features*

Other relevant features relating to the RIISOs as set forth in the various treaty agreements included the selection process for the head of the organisation (Director General for ITSO and IMSO and Executive Secretary for EUTELSAT IGO), term limits, physical location and staffing. Generally, the selection process mirrored the one that was in place prior to privatisation, with nominations made by the members and a voting process at a formal meeting. In both ITSO and IMSO, there are term limits of eight years (nominally two four-year terms), although it does not appear as though the Executive Secretary for EUTELSAT IGO is term limited.[55] For all three organisations, they are physically located at the same premises as the principal place of business of the privatised entity,[56] although this could change

[54] Art. 4, ITSO PSA.

[55] Art. X(b)(iii), ITSO Agreement; Art. 12, IMSO Convention; Art. X, EUTELSAT IGO Amended Convention.

[56] In the case of IMSO and EUTELSAT IGO, the principal place of business of Inmarsat and Eutelsat was located in the same jurisdiction in which the privatised entity was incorporated, which was the United Kingdom and France respectively. In the case of ITSO,

at some future time. Staffing is largely at the discretion of the Director General, although budgetary considerations tend to be the primary determiner of the extent of staffing. The full-time staffs of each organisation have remained small, with normally no more than a half a dozen full time staff and in the case of the EUTELSAT IGO, it has tended to be a single full time professional.

All three treaties also contain standard provisions relating to existence of juridical personality (including ability to contract, acquire and dispose of property and be a party to legal proceedings outside the scope of any privileges and immunities); delineation of functional responsibilities of the various organs (Assembly of Parties and Executive Organ or Secretariat headed by the Director General or Executive Secretary); arrangements dealing with the organisation's headquarters; means of accession or withdrawal; amendment; settlement of disputes among member Parties; entry into force; and identity and role of the Depositary.

One specific feature that was unique to IMSO concerned intellectual property rights in the name and logo. While in the case of both the privatised Intelsat and Eutelsat, these intellectual property interests were directly transferred to the privatised company, in the case of IMSO, the intellectual property rights related to both the name Inmarsat and the logo stayed with IMSO, although the privatised company was issued a royalty-free license to use the name and logo.[57] By the time of the INTELSAT and EUTELSAT privatisations, it would appear that the Parties had second thoughts about retention of intellectual property rights by the RIISO, which is the most likely reason for this change.

---

the principal place of business of Intelsat was in the United States, which was where ITSO was also located. However, this is not the same country as the country of incorporation, as Intelsat at the time of privatisation was incorporated in Bermuda, although in 2009 Intelsat's country of incorporation was changed to Luxembourg.

57 INMARSAT Assembly of Parties, *Report of the Twelfth Session of the INMARSAT Assembly*, Assembly/12/20, at para. 8.2.3 (5 May 1998): 18–19. Specifically the Assembly decided that the intergovernmental organisation would retain ownership of the Inmarsat name and logo; the use of the Inmarsat name and logo would be leased to the privatised company in perpetuity without charge and the privatised company would have unlimited rights to sub-license other entities to use the Inmarsat name and logo; the intergovernmental organisation would not allow any other entity to use the Inmarsat name and logo; and, in the event the intergovernmental organisation ceased to exist, at that time ownership of the Inmarsat name and logo would then pass to the privatised company.

## 4. *Significant Developments Following Privatisation*

In the case of both ITSO and EUTELSAT IGO, the most significant developments post-privatisation have been in relation to the influx of private equity investment into the privatised entities over which they have supervisory responsibilities. In the case of Intelsat, this also had a number of consequential impacts, specifically by resulting in the forced divestiture of the investment interests held by all former Signatories as a result of the private equity purchase of the privatised Intelsat, including all countries that fell into the category of LCO customers. In the case of both Intelsat and Eutelsat, it also resulted in the accumulation of very significant debt levels associated with the transactions, raising questions about the future viability of the organisations and the implications for the interests that the RIISOs were intended to safeguard in the event of a possible bankruptcy or insolvency. While the privatised Inmarsat also experienced a private equity buyout, this development appears to have been less of a concern for IMSO than for the other two RIISOs. Rather, in the case of IMSO, the most significant development post-privatisation related to the fairly ambitious expansion of the scope of responsibilities it was able to effectuate.

### 4.1. *ITSO-Related Developments*

As a result of two sequential private equity buyouts as well as Intelsat's merger with PanAmSat, which effectively resulted in the buyout of PanAmSat's private equity owners as well, Intelsat today has accumulated more than $15 billion in debt.[58] While its business prospects have remained strong, notwithstanding the global economic turndown, this debt level has certainly raised concerns about what might happen to the ITSO PSA's ongoing viability should an insolvency event occur. A similar concern arose as to what might happen to those satellite assets and associated orbital slots, which comprise the Common Heritage assets, under such circumstances. Questions posed included whether a bankruptcy court could invalidate the ITSO PSA, or whether the Common Heritage assets could be dispersed in such a way as to vitiate the possibility of future adherence to the Core Principles by one or more successor entities.

---

58 See Intelsat, *News Release, Intelsat Reports Full Year 2010 Results*, 11 March 2011. Available: http://www.intelsat.com/_files/investors/financial/2011/Q4_2010ER.pdf.

To address these concerns, ITSO resorted to two measures. The first was to seek to have the licensing process at the US Federal Communications Commission (FCC) modified in such a way that the licensing documentation referring to the Common Heritage assets issued by the FCC contained certain provisions that would protect those assets in the event of a bankruptcy or actions undertaken by Intelsat to dispose of them in a manner that dissipated the Common Heritage' overall value. ITSO originally sought to raise these concerns in connection with the proposed merger between Intelsat and PanAmSat.[59] At that time, the FCC indicated that was the wrong forum for their consideration, but pointed out that license modifications of the type that ITSO sought could be effectuated through a petition process under Section 316 of the Communications Act of 1934 (Communications Act),[60] and ITSO subsequently took up the FCC's suggestion in this regard. Specifically, ITSO requested the imposition of the following three conditions on all of the Common Heritage satellites that were now licensed by the FCC:

1. Ensuring that the Commission's licenses to Intelsat are linked to the Core Principles.
2. Ensuring that any successor to Intelsat, or other satellite operator that uses the Parties' Common Heritage assets, is bound by the Core Principles in the ITSO Agreement through the execution of a public services agreement with ITSO.
3. Requiring that Intelsat place a lien, letter of credit, third party guarantee or other legal instrument on certain satellites in order to provide bankruptcy protection to ensure the fulfillment of the "Core Principles" of the ITSO Treaty Agreement, including global connectivity, global coverage, nondiscriminatory access and protection of lifeline connectivity obligation (LCO) contracts. This protection would include the replacement of a sufficient number of satellites for the ongoing achievement of these goals.[61]

---

59 Comments of the International Telecommunications Satellite Organisation (ITSO) (filed 14 November 2004), submitted in Constellation, LLC et al., Consolidated Application for Authority to Transfer Control of PanAmSat Licensee Corp. and PanAmSat H-2 Licensee Corp., IB Docket No. 05-2.

60 Constellation, LLC, et al., Consolidated Application for Authority to Transfer Control of PanAmSat Licensee Corp. and PanAmSat H-2 Licensee Corp., IB Docket No. 05-290, Memorandum Opinion and Order, FCC 06-85, 21 FCC Rcd 7368, at para. 65 (2006): 7402.

61 Petition of the International Telecommunications Satellite Organisation (ITSO), IB Docket No. 06-137 (filed 10 July 2006).

Although ITSO's petition was initially opposed by Intelsat, subsequent correspondence from the US State Department[62] concurring with the appropriateness of the first two conditions sought by ITSO (but not the third), coupled with Intelsat's need to secure FCC approval of a proposed second private equity buyout, resulted in Intelsat's reassessment of its position and a willingness to accept the terms of Section 316 Petition submitted by ITSO as had been endorsed by the State Department.[63] In light of the State Department letter, ITSO also relented in terms of its insistence on the third condition. As a result, the FCC in 2008 imposed specific conditions in the licenses for all orbital locations transferred to the United States as part of the privatisation process indicating that these orbital resources could not be transferred to any other operator unless that operator first entered into a PSA with ITSO.[64]

While this development provided a degree of additional protection, ITSO remained concerned that, under the wording of the ITSO Agreement, in the event that Common Heritage assets were no longer used or needed by Intelsat, the United States, as Notifying Administration under Article XII(c)(ii) of the ITSO Agreement, would be obligated to cancel these orbital registrations under the procedures of the ITU without the possibility of any consideration being given to whether alternate arrangements could be considered, which would be consistent with the broader objectives embodied in the Core Principles. As such, this would result in the loss of a portion of the Common Heritage assets, which could have a serious impact on Intelsat's continued ability to meet the Core Principles. ITSO thought it would be preferable to at least preserve the possibility to determine

---

62 Letter from Ambassador David A. Gross, United States Coordinator, International Communications and Information Policy, US Department of State, to The Honorable Kevin J. Martin, Chairman, Federal Communications Commission, IB Docket No. 06-137 (15 March 2007): 1, 3–4 (recommending conditions that (1) explicitly obligate Intelsat to remain a signatory to the Public Services Agreement between Intelsat and ITSO approved by the ITSO Twenty-fifth Assembly of Parties and (2) provide, for licensing purposes, that no entity can be considered a successor-in-interest to Intelsat under the ITSO Agreement unless the entity has undertaken to perform the obligations of the Public Services Agreement).

63 Letter to The Honorable Kevin J. Martin, Chairman, Federal Communications Commission, from Ahmed Toumi, Director General & Chief Executive Officer, ITSO and Phillip L. Spector, Secretary, Intelsat North America L.L.C., IB Docket No. 06-137 (filed 19 October 2007).

64 Petition of the International Telecommunications Satellite Organisation (ITSO), IB Docket No. 06-137, Order of Modification, 23 FCC Rcd 2764 (2008).

whether another operator would be willing to enter into a type of PSA with it similar to the ITSO PSA with Intelsat, as a means of preserving the viability of these orbital resources as part of the Common Heritage. To that end, the Party of Colombia proposed an amendment at the Thirty-First (Extraordinary) Meeting of the ITSO Assembly of Parties in March 2007 that would revise the wording of Article XII(c)(ii) to allow for such a possibility. While certain Parties, including the United States, opposed the amendment, it was passed by the requisite two-thirds majority, although it will not enter into force until formal ratification by at least two thirds of the member states has occurred, which is a fairly cumbersome and time-consuming process. The text of the amendment as adopted by the Assembly reads as follows:

> In the event that the Company, or any future entity using the Common Heritage frequency assignments, waives such frequency assignment(s), uses such assignment(s) in ways other than those set forth in this Agreement, or declares bankruptcy, the Notifying Administrations shall authorise the use of such frequency assignment(s) only by entities that have signed a public services agreement, which will enable ITSO to ensure that the selected entities fulfill the Core Principles.[65]

To date approximately 70 countries have submitted their formal ratification of the proposed amendment, with 99 needed for it to enter into force. And while a significant degree of protection has been already made available by the grant of the Section 316 Petition, ITSO believes that the protection will be substantially more complete if the amendment is ultimately ratified.

### 4.2. *EUTELSAT IGO-Related Developments*

In the case of the EUTELSAT IGO, the private equity infusion, which occurred on a phased basis rather than in a single transaction, raised similar concerns, particularly as regards the continued commitment of the privatised Eutelsat to adhere to the principal objectives as well as to preserve the essential European character of the organisation. Indeed, as a result of those concerns, the EUTELSAT IGO Arrangement was amended as a result of negotiations between EUTELSAT IGO and Eutelsat that accorded the EUTELSAT IGO Executive Secretary status as an observer seat on the board

---

65 ITSO Assembly of Parties, *Record of Decisions of the Thirty-First (Extraordinary) Meeting*, AP 31-3E, at para 9 (6 April 2007): 4.

of the privatised company and the ability to weigh in on certain activities before their consummation. Additionally, through execution of a separate Letter Agreement, Eutelsat agreed to limit dividend distribution in excess of the amount of Eutelsat S.A.'s annual income or which would raise the Eutelsat S.A.'s ratio of net debt to EBITDA to more than 3.75 to one and to take appropriate measures to minimize the risk of a potential default, by committing, among other things, to maintain a consolidated debt level consistent with market practices and a minimum amount of shareholder equity consistent with sound financial management practices.[66]

These concerns first began to surface in 2004, prompting the EUTELSAT IGO Assembly of Parties to adopt a far reaching resolution calling upon the privatised Eutelsat not to change its Articles of Association in any way that would remove the protections that had been previously accorded to minority shareholders and that would require the privatised Eutelsat to notify the EUTELSAT IGO of any major changes in its Articles of Association. The resolution further requested the Executive Secretary of the EUTELSAT IGO to enter into negotiations with the privatised Eutelsat in order to define new ways of supervising Eutelsat including changes to the EUTELSAT IGO Arrangement. This was accompanied by the fairly explicit threat that, if the privatised Eutelsat was not agreeable to these measures being undertaken, or if the company precipitously increased its debt levels, such actions could be deemed as calling into question the company's ability to observe the Basic Principles.[67] Under the terms of the EUTELSAT IGO Amended Convention, any such failure to adhere to the Basic Principles could imperil Eutelsat's continued ability to utilize the frequency rights that had been retained within the EUTELSAT IGO.

This threat proved quite effective and resulted in the formal amendment of the Arrangement on December 10, 2004, in which a number of significant concessions were agreed to by the privatised Eutelsat. These included agreeing to provide 60 days notice of any proposal to change its operational, technical, commercial or financial policies that could materially affect its observance of the Basic Principles and to take into account

[66] Eutelsat, *Offering Memorandum*, Eutelsat Communications (29 November 2005): 23–24. Available: http://www.eutelsat.com/investors/pdf/regulated-info/IPO-Final-OM.pdf.

[67] EUTELSAT IGO Assembly of Parties, AP-33-03 (6–8 April 2005) (Attachment 1, Resolution of the 32nd Meeting of the EUTELSAT IGO Assembly of Parties on the Proposed Evolution of the Articles of Association of Eutelsat S.A. and the Adaptation of Its Warranties and Undertaking with Respect to the Basic Principles, 22 June 2004).

any request for modification submitted to it by EUTELSAT IGO within a specified time period of the relevant notification. Furthermore, Eutelsat agreed that no resolution for voluntary liquidation of Eutelsat for the purpose of restructuring or amalgamation shall be presented without the prior written consent of the EUTELSAT IGO. Eutelsat also committed that it would conduct an Initial Public Offering (IPO), subject to favourable market conditions, of 25 per cent of its shares by no later than December 7, 2006.[68] The IPO in fact occurred on December 2, 2005.

The most significant concession made, however, related to the future role that the Executive Secretary would play with respect to the Eutelsat Board of Directors. It was agreed that the Executive Secretary must be named a censor (observer) to the privatised Eutelsat's Board of Directors. In this capacity, the Executive Secretary would be able to attend and participate and have his or her interventions recorded in the minutes of all meetings of the Board, although not be able to cast a vote, and would have access to all information about the company under the same conditions as those applying to other members of the Board of Directors. The initial term would be for a three year period. Further, it would be renewable, with provision made for earlier termination in the event that a public offering occurred or that stock exchange authorities or the underwriting banks expressed the view that the continued role of the Executive Secretary as an official observer would have a significant negative impact on the public offering.[69]

The revised terms of the EUTELSAT IGO Arrangement were further modified the following year as a result of a separate Letter Agreement entered into on September 2, 2005, between Eutelsat Communications, a holding company that had been established by the private equity investors in the privatised Eutelsat in which their holdings in the privatised Eutelsat had been consolidated. With that Letter Agreement, the entity that would conduct the IPO would be Eutelsat Communications, rather than the privatised entity (Eutelsat S.A.), which would in effect be operated as a subsidiary of Eutelsat Communications. Moreover, Eutelsat Communications agreed to a number of undertakings that to some extent mirrored

---

68 Eutelsat, *Offering Memorandum*, Eutelsat Communications (29 November 2005): 66. Available: http://www.eutelsat.com/investors/pdf/regulated-info/IPO-Final-OM.pdf.

69 EUTELSAT IGO Assembly of Parties, Amendment No. 2 to the Arrangement Agreement Letter with a Group of Financial Shareholders, AP-33-7 (6–8 April 2005) (Attachment 1, Amendment No. 2 to the Arrangement between the European Telecommunications Satellite Organisation (EUTELSAT) and Eutelsat S.A.).

or exceeded the prior commitments that had been made by Eutelsat S.A., including the following:

- give the person holding the position of Executive Secretary of the EUTELSAT IGO a seat as a censor on the board of directors of Eutelsat Communications, beginning at the initial public offering of the latter;
- ensure that Eutelsat S.A. is at all times able to honor its commitments under the Arrangement and not take any decision likely to lead to any failure by Eutelsat S.A. to meet such obligations;
- in all cases, and without this being interpreted in any way as an exception to or weakening of the obligation referred to in the preceding paragraph, inform the Executive Secretary, in his/her position as a censor, of any decision approved by Eutelsat Communications likely to affect compliance with the Basic Principles by Eutelsat S.A. and to provide him/her with all relevant information on this matter;
- inform the EUTELSAT IGO, in the person of its Executive Secretary, of the crossing of any threshold, set by law or its bylaws, brought to its attention by a shareholder;
- not propose or approve any resolution for the distribution of dividends by Eutelsat S.A. exceeding the amount of Eutelsat S.A.'s annual net income or the sum of the annual net income and any annual net income allocated to reserves of Eutelsat S.A. or which would have the effect of increasing Eutelsat S.A.'s ratio of net debt to EBITDA to more than 3.75 to one, it being understood that this ratio will not be deemed to be exceeded as a result of an external growth transaction and that the definition of dividend is that in Article L.232-12 of the French Commercial Code;
- take all measures to ensure that the obligations undertaken or to be undertaken in the future by Eutelsat Communications, particularly for its current or future financing needs, do not in any way cause Eutelsat S.A. to default on its own financings, unless such obligations of Eutelsat Communications were also assumed in Eutelsat S.A.'s direct interest;
- maintain the Group's consolidated debt at a level which is not inconsistent with market practices and sound management of the Group; and
- keep a minimum amount of shareholders' equity in Eutelsat S.A. consistent with sound financial management of Eutelsat S.A. and allowing it to maintain its ability to comply with the Basic Principles.[70]

---

70 Eutelsat, *Offering Memorandum*, Eutelsat Communications (29 November 2005): 67. Available: http://www.eutelsat.com/investors/pdf/regulated-info/IPO-Final-OM.pdf.

The Letter Agreement was to take effect following the official listing of the shares of Eutelsat Communications on the Eurolist by Euronext Paris, which occurred in December 2005. The Letter Agreement will continue in force until the expiration of the Arrangement or upon the joint agreement of the parties thereto. Should controlling interest of Eutelsat S.A. be sold at any time by Eutelsat Communications, the buyer is to be made aware of the terms of the Letter Agreement, and such sale would not relieve Eutelsat Communications of its obligations under the Letter Agreement.[71]

### 4.3. *IMSO-Related Developments*

In the case of IMSO, the most significant post-privatisation developments have occurred in connection with efforts, largely successful, to expand the mandate of the organisation beyond that as defined at the time of privatisation in 1999, as well as to provide a distinct financing mechanism for at least some of these additional responsibilities. In September 2006, the IMSO Assembly of Parties approved (and then provisionally applied as of 2008) various amendments to the IMSO Convention. The aim of these amendments was to extend the oversight functions of IMSO to all mobile satellite service providers of GMDSS in the future and to permit IMSO to coordinate the establishment and operation of the international system for the LRIT worldwide. The activities that IMSO would undertake with respect to the LRIT function would not be funded by the same means as the GMDSS oversight functions, but instead would be funded through a series of separate agreements entered into with various national Administrations on whose behalf this function would be performed.[72]

The genesis for all these amendments appeared to originate within the IMO in the first instance, which had determined that robust intergovernmental oversight is necessary and should extend to all existing (Inmarsat Ltd) and new potential providers of GMDSS services in the future, and IMSO should be the oversight body on behalf of the international maritime community. Based on this, the Maritime Safety Committee (MSC) of the

71 Ibid.

72 IMSO Assembly of Parties, *Record of Decisions of the Eighteenth Session of the Assembly of the International Mobile Satellite Organisation*, ASSEMBLY/18/16/Corrected, at paras. 4.1–4.2 (30 October 2006): 4.

IMO invited IMSO to submit a proposal to effectuate this, which IMSO did in the form of an amendment to the IMSO Agreement.[73]

The amendment relating to LRIT similarly derived from decisions taken within the MSC of the IMO, which adopted in 2006, effective in 2009, a series of mandatory position reporting requirements applicable to certain categories of vessels, including passenger ships, cargo ships of 300 gross tonnage and upwards, and mobile offshore drilling units. Such vessels would be required to automatically transmit their identity and position with a date and time marker at regular six-hour intervals. At the invitation of the MSC, IMSO was formally designated as the LRIT Coordinator and charged with taking appropriate action to ensure the timely implementation of the LRIT System.[74]

As such, IMSO has been able to broadly recast its public service oversight to include not only GMDSS, but also distress alerting, search and rescue co-ordinating communications, maritime safety information broadcasts, general communications, as well as aeronautical safety services through compliance with the Standards and Recommended Practices established by the International Civil Aviation Organisation (ICAO). This does not diminish the importance of the GMDSS oversight function, but has given IMSO multiple rationales for continued existence.

## 5. *The Future*

While all three organisations face some uncertainties as to their future, each has sought to position itself as strongly as possible in preserving its future role. The most dramatic steps in this regard are those that have been undertaken by IMSO, which has successfully been able to expand its mission, such that the likelihood of it being phased out in the near term by the satisfaction of the conditions spelled out in the IMSO Agreement are now substantially diminished. More importantly, even if those conditions

73 Maritime Safety Committee, International Maritime Organisation, Report of the Maritime Safety Committee on Its Seventy-Seventh Session, MSC 77/26, at paras. 10.39–40 (10 June 2003): 79. Available: http://legacy.sname.org/committees/tech_ops/O44/imo/msc/77-26.pdf.

74 IMSO Assembly of Parties, *Long Range Identification and Tracking of Ships (LRIT) Background*, ASSEMBLY/20/7.1, Section 4.1 (25 June 2008): 1–2. Available: http://2001-2009.state.gov/documents/organisation/108683.pdf.

were satisfied, with the separate funding mechanism and new responsibilities now in place, IMSO would be able to continue.

In the case of EUTELSAT IGO, its retention of the orbital resources provides it with a rationale for continued existence, notwithstanding its more limited set of responsibilities. By using its dominion over these resources for maximum leverage, it has been able to expand the scope of its authority over the privatised Eutelsat and its publicly traded parent company, Eutelsat Communications, in a way that should extent its longevity for the foreseeable future.

In the case of ITSO, given that the decision has now been taken that the organisation will not be terminated at the end of twelve years after privatisation, and the question of termination has now been deferred until 2021, the main concern relates to ITSO's future funding. While funding provisions were secure through year twelve, the ITSO PSA does obligate Intelsat to continue to provide funding thereafter, although the maximum amount of such funding on an annual basis is capped. Moreover, the exact way in which the determination of the actual amount of annual funding to be provided was left somewhat nebulous, essentially dependent upon good faith negotiations between the Director General and the CEO of the privatised company.

One other development that may augur well for the continued existence of all three RIISOs has been their recognition of the common interests between them. Indeed, over the past few years, the three RIISO have entered into a formal Tripartite Memorandum of Understanding, executed in December 2009, which is intended to serve as a means for a formal sharing of information on activities of mutual interest, and for observer status at meetings of each of their respective Assemblies of Parties.[75] Starting in 2008, the three organisations have also embarked upon a series of tripartite meetings, at which issues of common interest are discussed relating to the respective oversight they provide of the three privatised entities.[76] At the 2009 meeting, for example, the heads of the three organisations reviewed

---

[75] ITSO Assembly of Parties, *Memorandum of Understanding Between ITSO, IMSO and EUTELSAT IGO*, AP-32-19 (6 September 2008).

[76] *See* EUTELSAT IGO, Press Release, Intergovernmental Satellite Organisation Tripartite Meeting (5 February 2008); Press Release, Second Intergovernmental Satellite Organisation Tripartite Meeting (4 December 2009); Press Release, Third Intergovernmental Satellite Organisation Tripartite Meeting (22 February 2011). Available: http://www.eutelsatigo.int/en/eutelsat.php?menu=9.

the instruments in place for monitoring the ability of the supervised companies to fulfil their obligations to the RIISOs in the longer term and consideration was also given to the role of the RIISOs in relation to the activities of other international organisations such as the ITU, the United Nations Committee on Peaceful Uses of Outer Space (COPUOS), the IMO, the ICAO and the European Union (EU).[77]

## 6. *Relations with other International Organisations*

As longstanding members of the international organisation community, the RIISOs have established and maintained a number of important relationships with many other international organisations.

While technically not formally part of the United Nations System of Organisations, all three international organisations, both prior and subsequent to privatisation, have strong ties with the United Nations and many other organisations that comprise the United Nations System. The bases upon which all three organisations were originally established were heavily tied to prior declarations of the United Nations, in particular Resolution 1721 (XVI) of the General Assembly, adopted in 1961, which was specifically referenced in the preambles to both the INTELSAT Agreement and INMARSAT Convention,[78] and which declared that "communication by means of satellites should be available to the nations of the world as soon as practicable on a global and non-discriminatory basis."[79] Moreover, all three treaty documents made specific reference to the Treaty on Principles Governing the Activities of States in the Exploration and Use of Outer Space, Including the Moon and Other Celestial Bodies.[80]

---

77 EUTELSAT IGO, Press Release, Second Intergovernmental Satellite Organisation Tripartite Meeting (4 December 2009).

78 Preamble, INTELSAT, INTELSAT Agreement; Preamble, INMARSAT Convention.

79 G.A. Res. 1721 (XVI), U.N. Doc. A/4987 (Dec. 20, 1961).

80 Preamble, INTELSAT Agreement; Preamble, INMARSAT Convention; Preamble, EUTELSAT Convention. The INTELSAT Agreement and INMARSAT Convention both went on to make specific reference to Art. I, Outer Space Treaty, which stated that outer space shall be used for the benefit and in the interests of all mankind; that provision was not expressly referenced in the EUTELSAT Convention, presumably on the basis that EUTELSAT was not intended, at least at the time of its creation, to function as a global satellite system in the same manner that INTELSAT and INMARSAT were conceived.

In addition, INTELSAT was obligated, consistent with the provisions of Resolution 1721 (XVI), to send to the Secretary General of the United Nations and the Specialized Agencies concerned, an annual report on the activities of INTELSAT.[81] A similar provision also appeared in the EUTELSAT Convention[82] although interestingly not in the INMARSAT Convention.

The INMARSAT Convention, however, contained an entirely separate article entitled "Relationship with Other International Organisations" which quite specifically dealt with how certain of those relationships were to be conducted:

> The Organisation shall co-operate with the United Nations and its bodies dealing with the Peaceful Uses of Outer Space and Ocean Areas, its Specialized Agencies, as well as other international organisations, on matters of common interest. In particular, the Organisation shall take into account the relevant international standards, regulations, resolutions, procedures and recommendations of the International Maritime Organisation and the International Civil Aviation Organisation. The Organisation shall observe the relevant provisions of the International Telecommunication Union Convention and regulations made thereunder, and shall in the design, development, construction and establishment of the INMARSAT space segment and of earth stations give due consideration to the relevant resolutions, recommendations and procedures of the organs of the International Telecommunications Union.[83]

In addition to the United Nations and its affiliated bodies, including the ITU and the IMO, the ISOs had a long history of entering into memoranda of understanding (MOU) with a multitude of other international organisations. Indeed, one of the specifically enumerated functions for the Assembly of Parties of each of the three ISOs, as exemplified by the wording of the INTELSAT Agreement, was to "decide upon questions concerning formal relationships between INTELSAT and States, whether Parties or not, or international organisations."[84] Prior to privatisation, the INTELSAT

---

81 Art. XXI(c), INTELSAT Agreement. The actual Specialized Agencies to which ITSO had sent the report include the ITU, the United Nations Educational, Scientific and Cultural Organisation (UNESCO); the World Intellectual Property Organisation (WIPO); the Universal Postal Union (UPU) and the World Bank.

82 Art. XXV(c), EUTELSAT Convention.

83 Art. 27, INMARSAT Convention.

84 Art. VII(c)(ix), INTELSAT Agreement. See Art. 12(f), INMARSAT Convention; Art. IX(a)(vi), EUTELSAT Convention.

Assembly of Parties had approved memoranda of understanding with a total of 15 other international organisations.[85] These agreements generally provided for voluntary information exchanges between the respective parties and a general spirit of cooperation, though their language tended to be quite general in nature and did not necessarily impose any particular substantive requirements on either party. Upon privatisation, these agreements remained with the RIISOs rather than being transferred to the privatised companies. Subsequent to privatisation, the RIISO's practice of entering into such MOUs has continued.[86]

The types of international organisations with which such agreements or relationships were established were somewhat variable, although for INMARSAT and EUTELSAT, there were certain obvious natural affinities. For IMSO, these affinities related to the maritime and aviation environment, and as such the strongest ties were to organisations such as the IMO, ICAO and Cospas-Sarsat (which has responsibility for the coordination of international search and rescue functions). Indeed, as an outgrowth of the IMO relationship, IMSO's expansion into the provision of LRIT services occurred. In the case of EUTELSAT IGO, the affinity would be with other

85 These MOUs were with the Intergovernmental Maritime Consultative Organisation/IMCO (approved at INTELSAT AP-2; the Asia Pacific Telecommunication unity (approved at INTELSAT AP-7); the International Maritime Satellite Organisation/INMARSAT (approved at INTELSAT AP-7); the European Telecommunications Satellite Organisation/EUTELSAT (approved at INTELSAT AP-10); the Intergovernmental Bureau for Informatics/IBI (approved at INTELSAT AP-10); the Arab Telecommunications Union/ATU (approved at INTELSAT AP-12); the Pan African Telecommunications Union/PATU (approved at INTELSAT AP-12); the Arab Satellite Communications Organisation/ARABSAT (approved at INTELSAT AP-12); the Central African Customs and Economic Union/UDEAC (approved at INTELSAT AP-15); the International Organisation of Space Communication/INTERSPUTNIK (approved at INTELSAT AP-18); the Caribbean Telecommunications Union/CTU (approved at INTELSAT AP-18); the Economic Community of West African States/ECOWAS (approved at INTELSAT AP-19); the Hispano-American Association of Research Centers and Telecommunications Corporations/AHCIET (approved at INTELSAT AP-19); the Central American Regional Technical Telecommunications Commission/COMTELCA (approved at INTELSAT AP-20) and the Commonwealth Telecommunications Organisation/CTO (approved at INTELSAT AP-20).

86 In the case of ITSO, the ITSO Assembly of Parties has approved three additional MOUs since privatisation – with the EUTELSAT IGO (approved at ITSO AP-28); with the Association of Telecommunications Companies of the Andean Community/ASETA (approved at ITSO AP-28); and jointly with the EUTELSAT IGO and IMSO (approved at ITSO AP-32).

pan-European organisations, such as the European Space Agency (ESA), the European Telecommunications Standards Institute (ETSI) and of course the EU.

On a more global level, the one other very strong and important relationship for all three organisations was with the ITU. This relationship touched not only on matters relating to orbital slots and radio frequency assignments, which are within the purview of the ITU Radiocommunications Bureau, but also common interests in many of the policy matters which arose within the ITU context. For ITSO and IMSO, this is now somewhat attenuated with respect to orbital slots and radio frequency assignments, since the orbital locations that each organisation had in its own right prior to privatisation were transferred to Notifying Administrations, which henceforth would also have responsibility for all future orbital slot and radio frequency filings submitted to the ITU on behalf of the privatised entities. In the case of ITSO, however, this transfer was not completely unconditional, given the emergence of the concept of the Common Heritage. With respect to EUTELSAT IGO, this connection to the ITU continues in a more direct fashion, because the orbital registrations held by EUTELSAT at the time of privatisation were not transferred to a Notifying Administration, but stayed with EUTELSAT IGO, although any new registrations or allotments undertaken following privatisation were to be handled by the Government of France as the Notifying Administration on behalf of the privatised Eutelsat.

In the case of the United Nations Committee on Peaceful Uses of Outer Space (COPUOS), the relationship is somewhat more informal, although the three RIISOs in 2009 specifically identified COPUOS as one of the organisations with which they desired to better coordinate their activities.[87] Moreover, as previously noted, the preamble to each of the treaty agreements does make explicit reference to the Outer Space Treaties and to the overriding objective of assuring the utilization of outer space for the benefit of and in the interests of all countries.

To date, none of the organisations appears to have established any formal working relationship with the World Trade Organisation (WTO)

---

[87] *See* EUTELSAT IGO, Press Release, Intergovernmental Satellite Organisation Tripartite Meeting (5 February 2008); Press Release, Second Intergovernmental Satellite Organisation Tripartite Meeting (4 December 2009); Press Release, Third Intergovernmental Satellite Organisation Tripartite Meeting (22 February 2011). Available: http://www.eutelsatigo.int/en/eutelsat.php?menu=9.

although they certainly have shared interest with the WTO in promoting broader market access arrangements, as reflected for example by the principles that underlie the separate Basic Agreement on Telecommunications. That document sets forth a set of regulatory principles that countries are expected to consider in the adoption of their schedules of basic telecommunications commitments as part of the General Agreement on Trade and Services formulated during the Uruguay Round of trade negotiations.[88] In addition to the ITU, the organisations can and do have dealings with various regional telecom bodies, such as the Inter-American Telecommunication Commission (CITEL) and Asia Pacific Economic Cooperation (APEC) forum.

One of the more interesting developments that has occurred concerns the United Nations Millennium Development Goals (MDGs), which were first articulated by the United Nations in the year 2000.[89] In 2010, based on concerns that broadband development was not proceeding at the desired pace, the United Nations established a separate Broadband Commission for Digital Development to address these concerns.[90] The Commission, which is jointly headed by Carlos Slim Helu and Rwandan President Paul Kagane, with the heads of the ITU and UNESCO serving as vice chairs, has been tasked with focusing attention on the shortfalls of achievement as well as the potential for enhanced development associated with more widely available broadband capabilities throughout the developing world. The Commission itself is made up of approximately 54 Commissioners, and the Director Generals of both ITSO and IMSO as well as the Executive Secretary of EUTELSAT IGO have all been appointed as Commissioners to this endeavour. The promotion of broadband services – indeed the articulation of broadband as a fundamental human right – has added a new dimension to the activities of the RIISOs. This is clearly reflective of the

---

88 World Trade Organisation, Agreement on Basic Telecommunications Services, the key elements of which are set forth in a Reference Paper, adopted on 24 April 1996, which contains a set of regulatory principles that countries are expected to consider in the adoption of their schedules of basic telecommunications commitments as part of the General Agreement on Trade and Services formulated as part of the Uruguay Round of trade negotiations. Available: http://www.wto.org/english/tratop_e/serv_e/telecom_e/tel23_e.htm.

89 G.A. Res. 55/2, U.N. Doc. A/RES/55/2 (Sept. 18, 2000).

90 Broadband Commission for Digital Development. Available: http://www.broadband-commission.org/.

important role that such individuals and the organisations they represent can play in the greater promotion of the availability of broadband around the globe, as a means of promoting various objectives consistent with the MDGs, including e-commerce, e-government and e-medicine.

## 7. *Assessment*

The experience leading up to the creation of the RIISOs and their subsequent mode of operation highlights certain of the difficulties associated with this type of restructuring. While the preparatory efforts leading up to privatisation of the ISOs were extraordinarily detailed and lengthy, and considerable effort was devoted to anticipating all possible contingencies, as experience demonstrates, that was impossible to achieve.

Clearly, the advent of private equity financing in the satellite sector, which has had a major impact on all three privatised entities and consequentially a major impact on the role and functioning of the RIISOs, had not been anticipated at all. Had it been so, it could have affected the manner in which the privatisations were effectuated and the conceptual underpinnings for the mission of the RIISOs, particularly with respect to possible safeguards for the investors from LCO or developing world countries to avoid being forced to sell their shares in the privatised company.

Another unexpected post-privatisation development was the willingness of the major initial shareholders, having shed their prior role and status as Signatories, to monetize their investments following privatisation. Indeed, while in one sense this seemed to be a predictable outcome given the views expressed by some more liberal governments, it was also clear that these governments did not necessarily fully expect this to happen. If they had, some of the more onerous measures insisted upon to protect against anticipated anticompetitive behaviours or to mandate substantial dilution of the original owners would have been unnecessary.

Probably the biggest difficulty facing the RIISOs was the constrained nature of the authority ultimately accorded to them, driven by the radically disparate perspectives on their future role and value held by the various Parties to the affected ISOs. For some Parties, the RIISOs were seen as necessarily evils, or the price that had to be paid to secure the necessary consensus in order for the privatisation of the ISOs to move forward. Those holding these views also wanted the RIISOs to have as limited amount of authority as possible, minimal amounts of resources available to them to discharge their duties, and relatively short timeframes in which they would be operational. For other Parties, the RIISOs were seen as an essential element of preserving the public service obligations that lie at the core of the ISOs, and the final safety net for dependent users whose interests were unlikely to be served by the dynamics of a purely commercial marketplace.

Given this disparity of views, the RIISOs encountered uncharted waters from the first day of their existence. The previous paradigm, in which the operational and regulatory functions were combined in a single entity provided a paucity of guidance as to how a purely supervisory ISO should function. Clearly, neither the necessary political consensus nor the successful outcome of the privatisation process could have been achieved in the absence of the RIISOs. At the same time, the rather limited mandate given to the RIISOs resulted in a fair degree of ambiguity as to exactly what role they were to perform and what vigour they were to bring to that role. The whole notion of a 'supervisory role', with authority not grounded on some broader mandate but arising principally from the terms of documents that were essentially contractual in nature (the PSAs or the Arrangement), made their proper functioning that much more difficult. This resulted in a high degree of frustration, not only on the parts of the RIISOs, but also the privatised companies as well as the various governments involved, particularly in their early days, regarding exactly how the RIISOS were to discharge their supervisory functions over the privatised entities.

Lastly, their unique role of overseeing specific competitive participants when other competing companies were not subject to such oversight ran the risk of continuing the specialness of the privatised entities that could affect the competitive balance of the marketplace, though whether it ultimately inured to the benefit or the detriment of the privatised entity could be open to further debate.

# Chapter Six

## Crossing a Rubycon? The International Legal Framework for ISOs – Before and After Privatisation

*F.G. von der Dunk*

### 1. *Introduction*

The present chapter analyses the activities of international satellite organisations (ISOs), former ISOs and private satellite operators from the perspective of the four principal international space law treaties,[1] consisting of the Outer Space Treaty,[2] the Rescue and Return Agreement,[3] the Liability Convention[4] and the Registration Convention.[5] In addition, it considers a United Nations Resolution dealing specifically with Direct Broadcasting by Satellite,[6] as it is one of the major categories of activities that international

---

1 The fifth international treaty usually considered part of the core of the *juris spatialis internationalis*, the Agreement Governing the Activities of States on the Moon and Other Celestial Bodies (hereinafter Moon Agreement), entered into force 11 July 1984, 1363 U.N.T.S. 3, 18 ILM 1434 (1979), is not relevant here.

2 Treaty on Principles Governing the Activities of States in the Exploration and Use of Outer Space, including the Moon and Other Celestial Bodies (hereinafter Outer Space Treaty), entered into force 10 Oct. 1967, 610 U.N.T.S. 205, 18 U.S.T. 2410, T.I.A.S. 6347.

3 Agreement on the Rescue of Astronauts, the Return of Astronauts and the Return of Objects Launched into Outer Space (hereinafter Rescue Agreement), entered into force 3 Dec. 1968, 672 U.N.T.S. 119, 19 U.S.T. 7570, T.I.A.S. 6599.

4 Convention on International Liability for Damage Caused by Space Objects (hereinafter Liability Convention), entered into force 1 Sept. 1972, 961 U.N.T.S. 187, 24 U.S.T. 2389, T.I.A.S. 7762.

5 Convention on Registration of Objects Launched into Outer Space (hereinafter Registration Convention), entered into force 15 Sept. 1976, 1023 U.N.T.S. 15, 28 U.S.T. 695, T.I.A.S. 8480.

6 Principles Governing the Use by States of Artificial Earth Satellites for International Direct Television Broadcasting, G.A. Res. 37/92, , at 39, U.N. Doc. A/AC.105/572/Rev.1 (Dec. 10, 1982).

satellite organisations such as INTELSAT and EUTELSAT have traditionally undertaken, as well as the International Telecommunication Union (ITU), which oversees the international regime developed to deal with frequency issues.[7]

Analysis here will not deal with the general substance of these agreements, but only with the specifics they may provide for dealing with ISOs, as opposed to sovereign states, and ISOs now under private ownership. In particular, it will focus on the extent to which the transition from ISOs to supervisory IGOs-*cum*-private operators marks a watershed. While politically and economically this transition has often been characterised as a kind of 'crossing of the Rubycon,' the question remains whether legally speaking that would be true as well.

## 2. *The Outer Space Treaty and Non-State Actors*

Turning to the 1967 Outer Space Treaty and its establishment of a baseline legal regime for outer space activities including satellite communications, this Treaty very clearly distinguishes between state actors and non-state actors. The second category can then be further subdivided into intergovernmental organisations and private operators.[8] It is by addressing the crossing of the boundary between those two subcategories that some light might be shed on the effects of the Treaty's regime on the privatisation process and *vice versa*, as well as on the way hybrid entities should be addressed from this perspective. The same general approach will also be followed with respect to the other treaties at issue, the aforementioned UN Resolution and the legal regime developed in the context of the ITU.

---

7 The ITU is the UN-affiliated global organisation dealing with international telecommunication issues, including those pertaining to satellite communications. It principally operates in accordance with the Constitution of the International Telecommunication Union (hereinafter ITU Constitution), entered into force 1 July 1994; 1825 U.N.T.S. 1; and the Convention of the International Telecommunication Union (hereinafter ITU Convention), entered into force 1 July 1994; 1825 U.N.T.S. 1. Both the ITU Constitution and the ITU Convention have been amended a few times since, notably in 1994 and 1998. See further *infra*, para. 7.2.

8 See on this issue in general Peter Malanczuk, "Actors: States, International Organisations, Private Entities," in *Outlook on Space Law over the Next 30 Years*, ed. Gabriel Lafferranderie and Daphne Crowther (Dordrecht: Kluwer Law International, 1997): 23–36.

## 2.1. *Intergovernmental Organisations and the Outer Space Treaty*

The focus of the present analysis is on the privatisation of a handful of organisations which in their original versions, as will be seen, were clearly intergovernmental in nature: INTELSAT,[9] INMARSAT[10] and EUTELSAT.[11] While to that extent the analysis of intergovernmental organisations would also apply to such entities as INTERSPUTNIK[12] and

9 INTELSAT was established on the basis of the Agreement Relating to the International Telecommunications Satellite Organisation (INTELSAT) (hereinafter INTELSAT Agreement), done 20 Aug. 1971, entered into force 12 Feb. 1973, as amended 17 Nov. 2000, amended version applied provisionally 18 July 2001, entered into force 30 Nov. 2004, 1220 U.N.T.S. 21, 23 U.S.T. 3813, T.I.A.S. 7532. See further Francis Lyall, *Law and Space Telecommunications* (Aldershot, Hants, Brookfield: Dartmouth, Gower, 1989): 74 ff., esp. 91–112; Francis Lyall and Paul B. Larsen, *Space Law – A Treatise* (Farnham: Ashgate, 2009), 325–37; Patrick A. Salin, *Satellite Communications Regulations in the Early 21st Century* (The Hague: Martinus Nijhoff, 2000): 101 ff., esp. 102–6; Ram S. Jakhu, "International Regulation of Satellite Telecommunications," in *Legal Aspects of Space Commercialization*, ed. Kunihiko Tatsuzawa (Tokyo: CSP Japan, 1992): 92–4.

10 INMARSAT was established on the basis of the Convention on the International Maritime Satellite Organisation (INMARSAT) (hereinafter INMARSAT Convention), done 3 Sept. 1976, entered into force 16 July 1979, as amended 1998, amended version applied provisionally 15 April 1999, entered into force 31 July 2001, 1143 U.N.T.S. 105, 31 U.S.T. 1, T.I.A.S. 9605. See further Lyall, *Law and Space Telecommunications*, 209 ff., esp. 219–28; Lyall and Larsen, 344–50; Salin, *Satellite Communications Regulations in the Early 21st Century*, 101, 120 ff., esp. 122–3; Jakhu, "International Regulation of Satellite Telecommunications," 94–5.

11 EUTELSAT was established on the basis of the Convention Establishing the European Telecommunications Satellite Organisation (EUTELSAT) (hereinafter EUTELSAT Convention), done 15 July 1982, entered into force 1 Sept. 1985, as amended 20 May 1999, amended version applied provisionally 2 July 2001, entered into force 28 Nov. 2002, Cmnd. 9069. See further Lyall, *Law and Space Telecommunications*, 264 ff., esp. 275–86; Lyall and Larsen, 356–60; Salin, *Satellite Communications Regulations in the Early 21st Century*, 365 ff.; Jakhu, "International Regulation of Satellite Telecommunications," 95–6.

12 Agreement on the Establishment of the "INTERSPUTNIK" International System and Organisation of Space Communications, entered into force 12 July 1972, 862 U.N.T.S. 3. See further Lyall, *Law and Space Telecommunications*, 296–303; Gennadi Zhukov and Yuri Kolosov, *International Space Law* (New York: Praeger, 1984): 109–20; Victor S. Veshchunov and Victoria D. Stovboun, "Intersputnik International Organisation of Space Communications: An Overview," *Journal of Space Law* 29 (2003): 121 ff.; Lyall and Larsen, 364–8.

ARABSAT,[13] the absence of full-fledged privatisation in the latter two cases causes them to be absent from this analysis. Although INTERSPUTNIK has more recently moved to a construct similar to the original structure of INTELSAT, INMARSAT and EUTELSAT, where the actual satellite operators conduct the day-to-day management of the system as opposed to the ruling body consisting of the states themselves, this level of 'hybridisation' would still not qualify INTERSPUTNIK as a private operator.[14]

### 2.1.1. *"Intergovernmental Organisations"*

By way of Articles VI and XIII, the Outer Space Treaty contains two relevant clauses specifying that its regime applies to international *intergovernmental* organisations ('IGOs,' which will be referred to as international *satellite* organisations or 'ISOs,' wherever, in view of the focus of the present analysis, IGOs undertaking satellite communication activities are specifically addressed), meaning that non-governmental international organisations ('NGOs') are not considered by those clauses.[15]

Both clauses furthermore deal with specific consequences following from the fact that relevant activities are carried out with the crucial involvement of an international intergovernmental organisation. Interestingly, IGOs are viewed as mere platforms for individual states to cooperate, rather than as separate entities: the relevant cases are described as "where ... [activities in the exploration and use of outer space] are carried on *within the framework* of international intergovernmental organisations."[16]

---

[13] Agreement of the Arab Corporation for Space Communications (ARABSAT), entered into force 15 July 1976, Space Law – Basic Legal Documents, C.VII.1. See further Lyall, *Law and Space Telecommunications*, 303–9; Adel A. Ziadat, "Arabsat: Regional Development in Satellite Communications: Lessons from the Arabsat Venture," *Zeitschrift für Luft- und Weltraumrecht* 37 (1988): 35 ff.; Lyall and Larsen, 375–7.

[14] See Protocol on the Amendments to the Agreement on the Establishment of the "INTERSPUTNIK" International System and Organisation of Space Communications, entered into force 4 Nov. 2002, Space Law – Basic Legal Documents, C.VIII.2. See further on INTERSPUTNIK's reform Veshchunov and Stovboun 122 ff.; Victor S. Veshchunov, "Reorganisation of INTERSPUTNIK," *Outer Space Z News* 2–2 (1999): 9–10; Lyall and Larsen, 368 ff.

[15] See e.g. Ulrike M. Bohlmann and Gisela Süss, "The status of international intergovernmental organisations under the UN Outer Space Treaty System," *Space Law – Newsletter of the International Bar Association Legal Practice Division* 10-1 (Oct. 2009): 8.

[16] Art. XIII, Outer Space Treaty; emphasis added. See further e.g. Ulrike M. Bohlmann and Gisela Süss, "Article XIII," in *Outer Space Treaty*, CoCoSL Vol. 1, ed. Stephan Hobe, Bernhard Schmidt-Tedd and Kai-Uwe Schrogl (Cologne: Carl Heymanns, 2009): 216–20.

This rather subordinated legal status of IGOs is furthermore confirmed by the second paragraph, where "practical questions arising in connection with activities carried on by international intergovernmental organisations ... shall be resolved by the States Parties to the Treaty either with the appropriate international organisation or with one or more States members of that international organisation, which are Parties to this Treaty."[17]

As such, states having a complaint about activities of an ISO are always entitled to refer to one or more of the member states of the latter, in case they do not wish to deal with the ISO itself. This absence of a measure of true legal independence for ISOs in the international community of space actors is largely the result of reluctance on the part of the Soviet Union at the time to grant IGOs such independent legal personality, and its insistence on 'classical' state sovereignty as the lynchpin for dealing with any legal issues pertinent to outer space.[18]

The same limited status for ISOs emanates from the other key article here, Article VI, which provides: "When activities are carried on in outer space ... by an international organisation, responsibility for compliance with this Treaty shall be borne both by the international organisation and by the States Parties to the Treaty participating in such organisation."[19] Responsibility of an IGO without the possibility to refer (also) to its individual member states, apparently, was not a feasible option.

---

17 Art. XIII, Outer Space Treaty.

18 See e.g. Paul G. Dembling and Daniel M. Arons, "The Evolution of the Outer Space Treaty," *Journal of Space Law* 33 (1967): 451–3; Bohlmann and Süss, "The status of international intergovernmental organisations under the UN Outer Space Treaty System," 8; Zhukov and Kolosov, esp. 4–17, 36; also Malanczuk, "Actors: States, International Organisations, Private Entities," 30; Stephan Hobe, *Die rechtlichen Rahmenbedingungen der wirtschaftlichen Nutzung des Weltraums* (Berlin: Duncker & Humblot, 1992): 72; in more general terms Malcolm N. Shaw, *International Law* (6th ed.) (Cambridge: Cambridge University Press, 2008): 31–6; Michael B. Akehurst, *A Modern Introduction to International Law* (5th ed.) (London: George Allen & Unwin, 1984): 16–9; Antonio Cassese, *International Law* (Oxford: Oxford University Press, 2001): 40. On the fundamental changes in this Soviet perspective due to 'glasnost' and 'perestroika,' see Vladimir S. Vereshchetin and Gregory V. Silvestrov, "Space Commercialization in the Soviet Union: Facts, Policy and Legal Issues," in *Legal Aspects of Space Commercialization*, ed. Kunihiko Tatsuzawa (Tokyo: CSP Japan, 1992): 32–40.

19 See further Michael Gerhard, "Article VI," in *Outer Space Treaty*, CoCoSL Vol. 1, ed. Stephan Hobe, Bernhard Schmidt-Tedd and Kai-Uwe Schrogl (Cologne: Carl Heymanns, 2009): 122–3.

In other words, the concept of an IGO, whether involved in satellite communications or in other space activities, is clearly delineated, if not indeed defined, by its having sovereign states as members. This conforms of course to the general public international law concept of an IGO.[20] As a consequence of such membership and the essentially public character of its tasks, responsibilities and activities, IGOs have an institutionalised system for participation of member state representatives,[21] may be granted secondary competencies to conclude international agreements of a treaty-like nature,[22] and enjoy certain functional immunities and privileges based on those of sovereign states and their diplomatic representations.[23]

Thus, in spite of their hybrid character (Public Telecommunication Operators and private operators, such as COMSAT or KDD, as signatories to the underlying operating agreements being in charge of the day-to-day running of the organisation as well as providing for the financing and the revenue-sharing mechanism), prior to their privatisation INTELSAT, INMARSAT and EUTELSAT were undoubtedly IGOs in the sense of the aforementioned Articles VI and XIII of the Outer Space Treaty. These ISOs were all composed of member states which were represented in the ultimate ruling body, and under whose control also the members of the chief governing body, consisting of representatives of the 'signatories,' were designated.[24] Also, they had Headquarters Agreements and related

20 See e.g. Ian Brownlie, *Principles of Public International Law* (7th ed.) (Oxford: Oxford University Press, 2008): 675–99, esp. 687–9; Peter Malanczuk, *Akehurst's Modern Introduction to International Law* (7th ed.) (London: Routledge, 1997): 92–3; Cassese, 69–71; Rebecca M.M. Wallace, *International Law* (3rd ed.) (London: Sweet & Maxwell, 1997): 68–9; Bohlmann and Süss, "The status of international intergovernmental organisations under the UN Outer Space Treaty System," 8.

21 Usually such member state participation is enshrined by means of their representation in key organs of the organisation labelled 'General Assembly,' 'Assembly of Parties,' 'Plenary Meeting,' 'Council' and suchlike.

22 Cf. also the Vienna Convention on the law of treaties between states and international organisations or between international organisations, Vienna, done 21 March 1986, *not yet entered into force*, Cm. 244, 25 I.L.M. 543 (1986).

23 Cf. e.g. the so-called Headquarters Agreements between IGOs and their host states, spelling out such immunities and privileges in detail (for the United Nations, e.g. the Agreement between the United Nations and the United States of America regarding the Headquarters of the United Nations, entered into force 21 Nov. 1947, 11 U.N.T.S. 11, 554 U.N.T.S. 308 (1966), 687 U.N.T.S. 408 (1969), and the general Convention on the Privileges and Immunities of the United Nations, entered into force 17 Sept. 1946, 1 U.N.T.S. 15).

24 For INTELSAT, cf. Art. I(a), (b), (f), (g), II, VI, VII (of which § (a) provides: "The Assembly of Parties shall be composed of all the Parties and shall be the principal organ of

Protocols on privileges and immunities in place with their respective host states: INTELSAT with the United States,[25] INMARSAT with the United Kingdom[26] and EUTELSAT with France.[27]

### 2.1.2. *The ISOs Post-Privatisation*

The privatisation process did not result in the complete elimination of IGO structures: INTELSAT as an ISO was replaced by the International Telecommunication Satellite Organisation or ITSO,[28] INMARSAT by the International Mobile Satellite Organisation or IMSO,[29] and EUTELSAT by

---

INTELSAT."), IX & X, INTELSAT Agreement. For INMARSAT, cf. Artt. 1 (a), (b), (c), 2, 4, 9–15, INMARSAT Convention. For EUTELSAT, cf. Artt. I(a), (b), (e), (f), II, VII–XII, EUTELSAT Convention.

25 Headquarters Agreement between the Government of the United States of America and the International Telecommunications Satellite Organisation, entered into force 24 Nov. 1976, T.I.A.S. 8542; resp. Protocol on INTELSAT Privileges, Exemptions and Immunities, entered into force 9 Oct. 1980, Cmnd. 8103.

26 Headquarters Agreement between the Government of the United Kingdom of Great Britain and Northern Ireland and the International Maritime Satellite Organisation, entered into force 25 Feb. 1980, 1203 U.N.T.S. 131; resp. Protocol on the Privileges and Immunities of the International Maritime Satellite Organisation (INMARSAT), entered into force 30 July 1983, 1328 U.N.T.S. 149.

27 Headquarters Agreement between EUTELSAT and the Government of the French Republic, done Nov. 15, 1985; resp. Protocol on the Privileges and Immunities of the European Telecommunications Satellite Organisation (EUTELSAT), entered into force 17 Aug. 1988, Cm. 1106, Cmnd. 305.

28 See Agreement Relating to the International Telecommunications Satellite Organisation (ITSO) (hereinafter ITSO Agreement), as amended 17 Nov. 2000, amended version applied provisionally 18 July 2001, entered into force 30 Nov. 2004, Cm. 5092. Further e.g. Sylvia Ospina, "International Satellite Organisations: Their Evolution from 'ISOs' to 'GCSs,'" in *Proceedings of the International Institute of Space Law 2010* (Reston: AIAA, 2011): 338–44; Lyall and Larsen, 337 ff.; Salin, *Satellite Communications Regulations in the Early 21st Century*, 468–72; Leo Millstein, "INTELSAT Restructuring," *Outer Space Z News* 2-2 (1999): 2–3; Francis Lyall, "On the Privatisation of INTELSAT," *Journal of Space Law* 28 (2000): 101–19.

29 See Convention on the International Mobile Satellite Organisation (hereinafter IMSO Convention), as amended 1998, amended version applied provisionally 15 April 1999, entered into force 31 July 2001, ATS 2001 No. 11. See further e.g. Ulrike M. Bohlmann, Kai-Uwe Schrogl and Ilaria Zilioli, "Report of the 'Project 2001' Working Group on Telecommunication," in *'Project 2001' – Legal Framework for the Commercial Use of Outer Space*, ed. Karl-Heinz Böckstiegel (Cologne: Carl Heymans, 2002): 219–20; Lyall and Larsen, 350 ff.; David Sagar, "INMARSAT: A New Beginning," *Outer Space Z News* 2-2 (1999): 6–8; David Sagar, "Inmarsat Since Privatisation," in *Proceedings of the Project 2001 Workshop on Telecommunication* (Cologne: Institute of Air and Space Law, 2000): 163–8; Salin, *Satellite Communications*

EUTELSAT IGO.[30] ITSO, IMSO and EUTELSAT IGO are still composed of member states as the ultimate bearers of competence to determine the legal structure, role and personality of the organisations;[31] and as such still undoubtedly qualify as IGOs. These three residual organisations, continue to enjoy functional privileges and immunities, especially in their respective host states of the United States, the United Kingdom and France.[32]

In fact, it is merely the scope of tasks and competencies of these ISOs which has been greatly reduced when compared to their predecessors. With a view to the application of the Outer Space Treaty, however, this does raise the question of whether these new IGOs are still materially impacted by the regime of the Outer Space Treaty, since Article VI thereof is concerned with 'activities in outer space' and Article XIII with 'activities in the use or exploration of outer space.'

ITSO's main purpose is now simply "*to ensure*, through the Public Services Agreement (PSA), *that the Company provides*, on a commercial basis, international public telecommunications services, in order to ensure performance of the Core Principles."[33] Article V of the ITSO Agreement

---

*Regulations in the Early 21st Century*, 472–4; Ospina, "International Satellite Organisations: Their Evolution from 'ISOs' to 'GCSs,'" 345 at n. 7.

30 See Convention Establishing the European Telecommunications Satellite Organisation (EUTELSAT) (hereinafter EUTELSAT Convention as amended), as amended 20 May 1999, amended version applied provisionally 2 July 2001, entered into force 28 Nov. 2002, Cm. 4572. Further e.g. Christian Roisse, "EUTELSAT Privatisation," *Outer Space Z News* 2-2 (1999): 4–5; Lyall and Larsen, 360 ff.; Ospina, "International Satellite Organisations: Their Evolution from 'ISOs' to 'GCSs,'" at n. 7.

31 Cf. e.g. Artt. I(a) & (p), II, IX, XI, XIV-XVIII, ITSO Agreement; Artt. 1(a) & (c), 6(1), 8, 12, 14, 16–18, IMSO Convention; and Artt. I(a) & (d), II(a), IV, VII, XI, XIII–XIV, XVI–XVIII, EUTELSAT Convention as amended; respectively. Further e.g. Ospina, "International Satellite Organisations: Their Evolution from 'ISOs' to 'GCSs,'" 338–9.

32 See Art. XIII(b) & (c), ITSO Agreement; Art. 9(5) & (6), IMSO Convention; and Art. XII(b) & (c), EUTELSAT Convention as amended; respectively; also Headquarters Agreement between the European Telecommunications Satellite Organisation (EUTELSAT) and the Government of the French Republic, done May 15, 2001, www.eutelsatigo.int/en/docs/HQ_agreement/pdf; see Lyall and Larsen, 362 incl. n. 157.

33 Art. III(a), ITSO Agreement; emphasis added. These Core Principles are further spelled out in Art. III(b). In addition, the functions of ITSO's main organ, the Assembly of Parties, in Art. IX(c) & (d), are essentially focused on that supervisory role. So also e.g. Ospina, "International Satellite Organisations: Their Evolution from 'ISOs' to 'GCSs,'" 338–42; also Francis Lyall, "The Protection of the Public Interest in the Light of the Commercialisation

further defines that purpose as 'supervision,' imposing a general obligation that "ITSO shall take all appropriate actions" to that end. Similarly, IMSO is "*to ensure* that the basic principles set forth in this Article *shall be observed by the Company*,"[34] and as for EUTELSAT IGO, "the primary purpose of EUTELSAT is *to ensure* that the Basic Principles set forth in this Article *are observed by the Company Eutelsat S.A.*"[35] These are all, clearly, not 'activities in outer space' in the sense of Article VI of the Outer Space Treaty.

By contrast, and as a consequence, in the case of ITSO "the international telecommunications satellite organisation's *space system is transferred*" to Intelsat,[36] in the case of IMSO it is Inmarsat which will henceforth operate the satellite system,[37] whereas in the case of EUTELSAT IGO, "Eutelsat S.A. will be established *to operate a satellite system and to provide satellite services* and for this purpose, EUTELSAT's assets and operational activities will be transferred to the Company Eutelsat S.A."[38]

---

and Privatisation of the Providers of International Satellite Telecommunications," in *Proceedings of the Forty-Seventh Colloquium on the Law of Outer Space* (Reston: AIAA, 2005): 442–4.

34 Art. 3, IMSO Convention; emphasis added. The phrase 'the Company' refers to Inmarsat. The remainder of Art. 3 on the substance of these 'basic principles,' as well as Art. 8 on the functions of the General Assembly of IMSO, further spell out the key elements of such supervisory function. See further e.g. Francis Lyall, "The Protection of the Public Interest in the Light of the Commercialisation and Privatisation of the Providers of International Satellite Telecommunications," 444–5.

35 Art. III(a), EUTELSAT Convention as amended; emphasis added. The remainder of Art. III(a) spells out the details of those Basic Principles. See further Art. IX for the functions of the Assembly of Parties in this regard.

36 Art. I(d), ITSO Agreement; emphasis added. See further Preamble, in particular its 5th para. Cf. also Isabel Polley, "Commentary Paper "International Satellite Service Providers," in *Proceedings of the Project 2001 Workshop on Telecommunication* (Cologne: Institute of Air and Space Law, 2000): 170–1; Ospina, "International Satellite Organisations: Their Evolution from 'ISOs' to 'GCSs,'" 339.

37 See Art. 1(b), IMSO Convention. See further Preamble, in particular its 6th para.; also e.g. Bohlmann, Schrogl and Zilioli, 219; Sagar, "Inmarsat Since Privatisation," 166.

38 Art. II(b) (i), EUTELSAT Convention as amended; emphasis added. See further Preamble, in particular its 5th para. stating "the will to transfer the operational activities and associated assets of EUTELSAT to a limited liability company to be established under a national jurisdiction, such company to operate on a sound economic and financial basis having regard to accepted commercial principles" as a major reason for privatisation.

### 2.2. *Private Operators and the Outer Space Treaty*

Whilst the Outer Space Treaty does not use the term 'private,' essentially because of the aforementioned political issues the Soviet Union had with acknowledging legal personality of any other entities than states, it does without doubt encompass private actors in outer space such as operators of telecommunication satellites in the broader term 'non-governmental entities' of Article VI,[39] as well as in the term 'nationals' of Article IX.[40]

#### 2.2.1. *"Authorisation and Continuing Supervision"*

Article VI provides for the main requirement applicable here to private satellite communication operators with the clause stating that "the activities of non-governmental entities in outer space, including the Moon and other celestial bodies, shall require authorisation and continuing supervision by the appropriate State Party to the Treaty." This clause is widely perceived to provide the most fundamental basis for national space laws and/or licensing systems for private space activities,[41] noting at the same time that the precise definition of 'the appropriate state' is still subject to discussion.[42]

---

[39] Art. VI, Outer Space Treaty, provides in relevant part: "States Parties to the Treaty shall bear international responsibility for national activities in outer space … carried on … by non-governmental entities."

[40] Art. IX, Outer Space Treaty, refers to "an activity or experiment planned by it or its nationals in outer space." See further e.g. Sergio Marchisio, "Article IX," in *Outer Space Treaty*, CoCoSL Vol. 1, ed. Stephan Hobe, Bernhard Schmidt-Tedd and Kai-Uwe Schrogl (Cologne: Carl Heymanns, 2009): 179–80.

[41] See the author's *Private Enterprise and Public Interest in the European 'Spacescape'* (Leiden: IIASL, 1998): esp. 17–22; Elisabeth Back-Impallomeni, "Article VI of the Outer Space Treaty," in *Proceedings of the United Nations/Republic of Korea Workshop on Space Law* (Vienna: UNOOSA, 2004): 75; Kunihiko Tatsuzawa, "Policy and Law in Space Commercialization," in *Legal Aspects of Space Commercialization*, ed. Kunihiko Tatsuzawa (Tokyo: CSP Japan, 1992): 19; René Oosterlinck, "Private Law Concepts in Space Law," in *Legal Aspects of Space Commercialization*, ed. Kunihiko Tatsuzawa (Tokyo: CSP Japan, 1992): 45–8; Gabriella Venturini, "Private Actors and Space Law: The Influence of Competition on Satellite Communications," in *Outlook on Space Law over the Next 30 Years*, ed. Gabriel Lafferranderie and Daphne Crowther (Dordrecht: Kluwer Law International, 1997): 59.

[42] Cf. e.g. Michael Gerhard, "The State of the Art and Recent Trends in the Development of National Space Law," in *Nationales Weltraumrecht/National Space Law*, ed. Christian Brünner and Edith Walter (2008): 65–6; Back-Impallomeni, "Article VI of the Outer Space Treaty," 75–6; Elisabeth Back-Impallomeni, "Necessities for the Development of National

Several authors in addition have claimed that, without such proper authorisation and continuing supervision, private activities in space would actually be prohibited, as far as international space law is concerned, since Article I of the Outer Space Treaty expressly limits the freedom to explore and use outer space to *states*.[43]

At the same time, Article VI determines that the state(s) whose national activities are concerned would remain responsible in any event, effectively comprising such activities if conducted by non-governmental entities on the same footing.[44] Thus, in case those activities would transgress the boundaries set by international (space) law, the state(s) concerned would have to be addressed for repairing such an internationally wrongful act regardless of whether and to what extent such authorisation and supervision had been provided. This, obviously, would include private operators of satellite systems for communication purposes.[45]

Furthermore, it has been a comprehensive and uniform general practice for states not to question whether private activities in outer space conducted by operators within the jurisdiction of other states have actually been licensed by those states, nor is there any international custom of international public notification of granting such licenses. The resulting general absence of protests from states to cases where other states might have allowed such private activities to go ahead without any specific license or authorisation constitutes a major argument for concluding that non-authorised and/or non-supervised activities do not violate any

---

Space Law," in *Nationales Weltraumrecht/National Space Law*, ed. Christian Brünner and Edith Walter (2008): 28–30; Karl-Heinz Böckstiegel, "The Term 'Appropriate State' in International Space Law," in *Proceedings of the Thirty-Seventh Colloquium on the Law of Outer Space* (Reston: AIAA, 1995): 77–9; Henri A. Wassenbergh, *Principles of Outer Space Law in Hindsight* (Dordrecht: Martinus Nijhoff, 1991): 24–9; the author's *Private Enterprise and Public Interest in the European 'Spacescape,'* 19–21.

43 See Manfred Lachs, *The Law of Outer Space – An Experience in Contemporary Law-Making* (Reprint Leiden: Brill, 2010): 114; Paul G. Dembling, "Treaty on Principles Governing the Activities of States in the Exploration and Use of Outer Space Including the Moon and Other Celestial Bodies," in *Manual on Space Law*, ed. Nandasiri Jasentuliyana and Roy S.K. Lee, Vol. I (Dobbs Ferry: Oceana Publications, Inc.; Alphen aan den Rijn: Sijthoff & Noordhoff, 1979): 17; Back-Impallomeni, "Article VI of the Outer Space Treaty," 73. Cf. also Tatsuzawa, 19; Back-Impallomeni, "Necessities for the Development of National Space Law," 29.

44 See also Lachs, 114; Lyall and Larsen, 66; Wassenbergh, 28; Gerhard, "Article VI," 112–4.

45 Cf. also e.g. Lyall and Larsen, 378–9.

rule of space law by the mere fact of such lack of authorisation and/or supervision.[46]

Indeed, it would seem impractical for other states to merely call a state to account for not having (properly) authorised and/or continuously supervised certain private space activities if such space activities would not actually have violated substantive obligations under space law. Hence, the more appropriate view would seem to be that the freedom for states to explore and use outer space subsumes the corresponding freedom for their private entities, even if not authorised and supervised; but under no circumstance would such an absence of authorisation and supervision detract from the state's responsibility for the private space activities at issue, and actually would enlarge the extent of the violation of relevant treaty obligations.[47] The absence of specific authorisation and/or supervision from such a perspective could be read as a blank approval by the state to conduct the activities concerned, since it will be held responsible regardless; it (presumably knowingly) accepts the risk that its responsibility will be invoked, yet has not chosen to develop and/or apply any specific authorisation and/or supervision regime.

It is once such activities come to be in violation of material rules of international (space) law that states concerned would invoke the responsibility of the relevant other state(s) in order to see the violation stopped and/or reparation provided. This key principle of international state responsibility goes back to the famous Chorzow Factory case of 1928, where the Permanent Court of International Justice stated that the fundamental legal consequence of state responsibility for an international wrongful act would be an obligation for the responsible state to provide for 'reparation,' such reparation usually taking the form of either *restitutio in integrum*, compensation (usually in case only of material harm) and/or satisfaction.[48]

---

46 Contra, however, e.g. again Dembling, "Treaty on Principles Governing the Activities of States in the Exploration and Use of Outer Space Including the Moon and Other Celestial Bodies," 17: 'it would appear that Article VI requires a certain minimum of licensing and enforced adherence to government-imposed regulations."

47 Cf. also Wassenbergh, 22 ff.; also Lyall and Larsen, 66.

48 Case concerning the factory at Chorzów (Merits) (Germany v. Poland), Permanent Court of International Justice, Sept. 13, 1928, P.C.I.J., Ser. A, No. 17; see further e.g. Shaw, 800–8, esp. 801; Brownlie, 434–40, esp. 435; Wallace, 175–6; Raymond Bender, *Space Transport Liability* (The Hague: Kluwer Law International, 1995): 284–9; the author's "Liability versus Responsibility in Space Law: Misconception or Misconstruction?," in *Proceedings of the Thirty-Fourth Colloquium on the Law of Outer Space* (Reston: AIAA, 1992): 363–71.

### 2.2.2. *"National Activities (in Outer Space)"*

The other major issue debated in the context of Article VI and its application to private actors concerns the scope of the concept of 'national activities (in outer space).' Without detailing the many discussions on the topic and the many definitions or circumscriptions offered,[49] the current status as regards the definition of 'national activities' lacks clarity on the international level. This leads those states that find themselves compelled to implement the authorisation and supervision clauses of Article VI by means of national laws and/or licensing systems to follow their own preference in applying such laws or systems to nationals (including companies with their respective nationality) and/or to anyone operating from their territory, and/or through still other channels for exercising jurisdiction.[50]

Article IX of the Outer Space Treaty refers to the more limited context where states are aware "that an activity or experiment planned by it or its nationals in outer space ... would cause potentially harmful interference with activities of other States Parties in the peaceful exploration and use of outer space," in which case such states "shall undertake appropriate international consultations before proceeding with any such activity or experiment." In other words, a specific obligation for states to consult with other states in certain scenarios involving their respective national companies (including in cases of satellite communication activities) is added to the all-encompassing obligation to be held accountable for the activities of those companies in case of transgressing the boundaries imposed by international (space) law in this respect.

### 2.2.3. *The Private Character of Intelsat, Inmarsat and Eutelsat and their Licensing*

There can be no doubt that Intelsat, Inmarsat and Eutelsat are private companies. They do not have a membership of sovereign states but are principally owned by shareholders (amongst which may be found some public

49 One might refer e.g. to the author's *Private Enterprise and Public Interest in the European 'Spacescape,'* 18–9; also Wassenbergh, 23 ff.

50 Cf. e.g. Malanczuk, *Akehurst's Modern Introduction to International Law*, 109–13; Brownlie, 299–308; Shaw, esp. 652–73; Wallace 111–9; also Gabriel Lafferranderie, "Jurisdiction and Control of Space Objects and the Case of an International Intergovernmental Organisation (ESA)," *Zeitschrift für Luft- und Weltraumrecht* 54 (2005): 230; on other bases accepted under international law for the exercise of sovereign jurisdiction.

authorities, but in an essentially private capacity), and they do not enjoy any of the financing or immunity-related privileges that IGOs can enjoy.[51] This means they are to be authorised and supervised, as far as Article VI is concerned, by "the appropriate State," and under Article IX by the state whose 'nationality' they carry (which under general international law used to refer to the state where the private operator is incorporated and headquartered).[52] Since in all three cases the state of registration is identical to the state of headquartering, Intelsat (headquartered in Washington at least for its administrative and operational purposes as was the old INTELSAT) qualifies as a US company, Inmarsat (headquartered in London as was the old INMARSAT) as a British one and Eutelsat (headquartered in Paris as was the old EUTELSAT) as a French one.[53] With respect to Intelsat, it should be added that essentially it was the administrative headquarters, including the satellite operational control functions, which remained in the United States,[54] as the formal incorporation of the holding company has recently moved to Luxembourg from Bermuda, a UK dependency where it was originally established.

The United States thus applies its Communications Act of 1934,[55] as amended by the Telecommunications Act of 1996, as the national act providing for authorisation and supervision in the sense of Article VI, to Intelsat. The Act provides for a licensing obligation of all persons using or

---

51 See e.g. Shaw, 775–6; Malanczuk, *Akehurst's Modern Introduction to International Law*, 127–8; Cassese, 72–4; also 91–7; Wallace, 133; Bohlmann and Süss, "The status of international intergovernmental organisations under the UN Outer Space Treaty System," 9.

52 Cf. Case Concerning the Barcelona Traction Light and Power Company, Limited (Second Phase) (Belgium v. Spain), International Court of Justice, 5 February 1970, I.C.J. Rep. 1970, 4, at 42, § 70.

53 Cf. e.g., for Inmarsat, Bohlmann, Schrogl and Zilioli, 219; Lyall and Larsen, 351; for Eutelsat, Lyall and Larsen, 360–1.

54 See e.g. available: http://en.wikipedia.org/wiki/Intelsat; http://www.intelsat.com/contact-us/corpo-rate/offices.asp; http://www.nytimes.com/2001/07/23/business/technology-satellite-com-pany-is-trying-life-on-its-own.html?scp=1&sq=collections%20intelsat%202001&st=cse; all accessed last 20 June 2012. Also e.g. Lyall and Larsen, 341. In addition to the United States, which was designated as the Notifying Administration for all existing satellites and satellite network filings operating in or involving C and Ku-band frequency assignments at the time of privatisation, the United Kingdom was designated as the Notifying Administration for various Ka-band and V-band satellite network filings that been previously been made on INTELSAT's behalf prior to privatisation. See Ospina, "International Satellite Organisations: Their Evolution from 'ISOs' to 'GCSs,'" 340.

55 Communications Act, June 19, 1934; 47 U.S.C. 151 (1988); 48 Stat. 1064.

operating "any apparatus for the transmission of … communications or signal by radio" from within the territorial or quasi-territorial jurisdiction of the United States.[56] The Federal Communications Commission (FCC), the US government agency which authorises and supervises US communications activities and operators, specifically declared in 1970 that the Communications Act was applicable to satellite communications.[57]

With reference to the two other countries formally involved in Intelsat, it should be pointed out that Luxembourg, the country of formal incorporation of the parent holding company,[58] does not have a national space law yet, which essentially makes application of US national law as per the above undisputed. In the case of the United Kingdom, which does have its Outer Space Act,[59] there are no actual satellites deployed at the orbital locations and operating in the frequency assignments for which the United Kingdom serves as the Notifying Administration at this time, leaving any potential role for the United Kingdom in this regard in a relatively passive state.[60]

In the case of Inmarsat, the United Kingdom does require Inmarsat to hold a license under the UK Outer Space Act, as it obliges any UK operator procuring the launch of a satellite, operating it or conducting any other activity in outer space to obtain a license in accordance with its terms.[61]

---

56 Sec. 301, Communications Act.

57 See Communications Satellite Facilities, *First Report and Order*, 22 FCC 2d 86 (1970), App. C, 1.

58 See available: http://en.wikipedia.org/wiki/Intelsat; http://www.intelsat.com/contact-us/corporate/offices.asp; http://www.nytimes.com/2001/07/23/business/technology-satellite-company-is-trying-life-on-its-own.html?scp=1&sq=collections%20intelsat%202001&st=cse; all accessed 20 June 2012.

59 Outer Space Act, July 18, 1986, 1986 Chapter 38, National Space Legislation of the World, Vol. I (2001), at 293, Space Law – Basic Legal Documents, E.I.

60 Cf. Ospina, "International Satellite Organisations: Their Evolution from 'ISOs' to 'GCSs,'" 346 at n. 15. It may be noted here, that Sec. 3(3), Outer Space Act, provides: "The Secretary of State may by order except other persons or activities from the requirement of a licence if he is satisfied that the requirement is not necessary to secure compliance with the international obligations of the United Kingdom." See further on the Outer Space Act, e.g. Irmgard Marboe and Florian Hafner, "Brief Overview over National Authorisation Mechanisms in Implementation of the UN International Space Treaties," in *National Space Legislation in Europe*, ed. Frans G. von der Dunk (Leiden: Brill, 2011): 35–6; Sa'id Mosteshar, "Regulation of Space Activities in the United Kingdom," in *National Regulation of Space Activities*, ed. Ram S. Jakhu (Dordrecht: Springer, 2010): 359–62. However, as noted, the United Kingdom does act as Notifying Administration with the ITU; see Ospina, "International Satellite Organisations: Their Evolution from 'ISOs' to 'GCSs,'" 340.

61 See Sec. 1, 2(1), Outer Space Act.

In addition, in view of the telecommunication activities conducted by its space system, Inmarsat is also required to comply with any relevant requirements under the UK Telecommunications Act; the mere existence of a license under the Outer Space Act does not take away such a requirement.[62]

As for France, the privatisation or restructuring of EUTELSAT, resulting in the establishment of the French company Eutelsat S.A., presented a major reason for recently enunciating a national space law where none had existed before.[63] The new Act obliges Eutelsat to hold a French license, as it requires "any ... juridical person carrying out a space operation" (which includes "ensuring the commanding of a space object during its journey in outer space"), if such operator is of French nationality, to be authorised by the French authorities under the Law.[64]

### 2.3. *The 'Boundary' Issue Under the Outer Space Treaty*

From the above analysis, the essential boundary between application of the Outer Space Treaty to ISOs and such application to private companies becomes clearly visible. The main difference between the two categories

---

62 Telecommunications Act, 1984 Chapter 12. See further Phillip J. Dann, "Law and Regulation of Satellite Communications in the United Kingdom," *Journal of Space Law* 20 (1992): 17–21; cf. also Sa'id Mosteshar, *European Community Telecommunications Regulation* (London, Boston, Norwell: Graham & Trotman/M. Nijhoff, Kluwer Academic Publishers, 1993): 8.

63 Law on space activities (*Loi relative aux opérations spatiales*), *Loi n° 2008-518*, June 3, 2008, 34 *Journal of Space Law* (2008): 453, unofficial translations 34 *Journal of Space Law* (2008): 453, *Nationales Weltraumrecht / National Space Law* (2008): 211. Cf. the presentation of Philippe Clerc on the 'French Implementation of Outer Space Treaty Art. VI under Space Operations Act June 3, 2008' at the Third Eilene M. Galloway Symposium on Critical Issues of Space Law, Washington, 11 December 2008. Available: http://rescommunis.files.wordpress.com/2008/12/3rd-galloway-clerc.pdf; Armel Kerrest de Rozavel and Frans G. von der Dunk, "Liability and Insurance in the Context of National Autorisation," in *National Space Legislation in Europe*, ed. Frans G. von der Dunk (Leiden: Brill, 2011): 150–61; Marboe and Hafner, 39–40; Philippe Clerc and François Cahuzac, "Advance in the Implementation of the French Space law on Space Operations in the Launcher Field," in *Proceedings of the International Institute of Space Law 2009* (Reston: AIAA, 2010): 400–6; Philippe Achilleas, "Regulation of Space Activities in France," in *National Regulation of Space Activities*, ed. Ram S. Jakhu (Dordrecht: Springer, 2010): 109–12.

64 Artt. 1(2) (& (3), 2(3), Law on space activities.

under the Treaty itself, it may be noted, consists in the requirements of authorisation and continuing supervision under Article VI and of consultation in cases of potentially hazardous activities under Article IX as far as the latter is concerned, although this may be more of a formal distinction. ISOs in practice will operate under some system of authorisation and possibly also supervision by their member states – only, in this case, through their governing multilateral treaties and statutes instead of through a single-state licensing process. Similarly, Articles VI and XIII jointly make clear that states are also required to consult in appropriate cases targeted by Article IX since ISOs are principally viewed as 'frameworks' for state activities, not as independent actors whose responsibilities are separate or to be separated from those of their member states.

In any privatisation process, as far as such a formal distinction goes, the Rubycon is crossed the moment the entities in question no longer consist of member states.[65] At that point in time, they come to fall under a single state's international responsibility as opposed to the joint responsibility of the member states – in the case of Intelsat, the United States; Inmarsat, the United Kingdom; and Eutelsat, France.

This single-state responsibility arises notwithstanding the continuing existence of ITSO, IMSO and EUTELSAT IGO respectively and their exercise of legally entrenched control and supervision activities over the three now private companies, as such control and supervision remain limited to a few explicitly delineated public interest aspects.

Thus, ITSO can "take all appropriate actions, including entering into the Public Services Agreement, to supervise the performance by the Company of the Core Principles;" the legally binding PSA with the company being the principal instrument for ITSO to make sure Intelsat will provide its services with "global connectivity and global coverage," continue to "serve its lifeline connectivity customers" and "provide non-discriminatory access to" Intelsat's satellite capacity.[66] Interestingly, at least for the first twelve years since privatisation was realised the costs involved in running ITSO will be

---

65 Cf. also e.g. the definition on Wikipedia: "Privatisation (…) is the incidence or process of transferring ownership of a business, enterprise, agency, public service or property from the public sector (the state or government) to the private sector (businesses that operate for a private profit) or to private non-profit organisations." Available: http://en.wikipedia.org/wiki/Privatisation.

66 Art. V, resp. Art. III(b), ITSO Agreement; see also Art. I(j).

met "by the retention of certain financial assets at the time of transfer of ITSO's space system to the Company."[67]

IMSO likewise has a Public Service Agreement in force with Inmarsat, spelling out the details of the obligations of the latter subject to supervision and control by the former.[68] The IMSO Assembly consequently is entitled "to take any steps or procedures necessary to ensure observance by the Company of the basic principles, as provided for in Article 4, including approval of the conclusion, modification and termination of the Public Services Agreement under Article 4(1)."[69] These basic principles oblige Inmarsat to ensure the continued provision of global maritime distress and safety satellite communications services, to provide services without discrimination on the basis of nationality, to act exclusively for peaceful purposes, to serve – at least in principle – all areas where there is a need for mobile satellite communications, giving due consideration to the rural and the remote areas of developing countries, and to operate in a manner consistent with fair competition.[70] Also IMSO is essentially funded by the company it is charged to supervise and control, at least to the extent of costs concerning the "establishment and operation of the Secretariat; the holding of Assembly sessions; and the implementation of any measures taken by the Organisation in accordance with Article 4 to ensure that the Company observes the basic principles."[71]

---

67 Art. VII(a), ITSO Agreement. After that twelve-year period, the funding of ITSO is to be arranged through the Public Services Agreement. Cf. also Ospina, "International Satellite Organisations: Their Evolution from 'ISOs' to 'GCSs,'" 340–4.

68 In its original version, this was the Public Services Agreement Between the International Mobile Satellite Organisation and Inmarsat One Limited and Inmarsat Two Company (1998), IMSO Docs. k:legalwpaes3338L.

69 Art. 8(b), IMSO Convention.

70 See Art. 3, IMSO Convention; also Art. 2, Public Services Agreement.

71 Art. 10(1), sub (a), (b) & (c), respectively, IMSO Convention; see also Art. 15, Public Services Agreement, detailing such obligations to the extent of requiring that "the Company shall pay to the Organisation the sum of 300,000 pounds sterling annually for operational expenses, which amount shall be amended on the first and each subsequent anniversary of the date of this Agreement to reflect the change, if any, in the published UK Retail Price Index (RPI) compounded annually from the date of this Agreement" (para. (3) (a)), and that "the Company shall also establish a contingency fund of 100,000 pounds sterling in the name of the Organisation to meet the Organisation's costs of arbitration or other legal proceedings in connection with enforcement proceedings taken by the Organisation under this Agreement" (para. (3) (d)).

Finally, EUTELSAT IGO once again has largely followed the path of ITSO, although here the Public Services Agreement has been labelled 'Arrangement,' defined as "the Arrangement between EUTELSAT and the Company Eutelsat S.A. having as its purposes to define the relationship between EUTELSAT and the Company Eutelsat S.A. and their respective obligations and in particular to provide the framework that will enable EUTELSAT to oversee and ensure the observance by the Company Eutelsat S.A. of the Basic Principles."[72] These Basic Principles pertain to universal service obligations, pan-European coverage by the satellite system, non-discrimination, fair competition, and due adherence to rights and obligations under international law such as the ITU Radio Regulations for the use of frequencies deriving from the operation of the space segment transferred from EUTELSAT IGO to Eutelsat.[73] EUTELSAT's main organ, the Assembly of Parties, consequently has the mandate *inter alia* "to consider and to review the activities of the Company Eutelsat S.A. which relate to the Basic Principles," "to ensure observance by the Company Eutelsat S.A. of the Basic Principles, in accordance with the Arrangement" and "to take the appropriate decisions in order to ensure continuity regarding rights and obligations under international law" as referred to before.[74] Also in the European context, the company basically has to pay: "The costs incurred in establishing and operating the Secretariat, including, but not limited to rent and associated costs of maintaining office premises, salaries and emoluments of staff, costs of organising and holding meetings of the Assembly of Parties, the costs of consultations between EUTELSAT and the Parties and other organisations and costs of applying measures taken by EUTELSAT under Article III to ensure that the Basic Principles are observed by the Company Eutelsat S.A. shall be borne by the Company Eutelsat S.A. in accordance with paragraph a) of Article V, within the relevant ceilings set forth in the Arrangement."[75]

Thus, it may be concluded that in each case crossing the Rubycon towards privatisation has resulted in the new operators now being subjected to formalised and explicit requirements as to their key international public duties, accompanied by substantial legal and jurisdictional tools for the revamped IGOs to ensure conformity with those requirements.

---

72 Art. I(l), EUTELSAT Convention as amended. Cf. also Art. II(d).

73 See Art. III, EUTELSAT Convention as amended.

74 Art. IX(a), (b) & (d), respectively, EUTELSAT Convention as amended.

75 Art. V(b), EUTELSAT Convention as amended.

By contrast, in the context of the ISOs prior to privatisation such public duties were considered to be sufficiently guaranteed by the control of the totality of their member states through the governing bodies.

The explicit character of those public duties and the need to preserve them in a privatised environment may well turn out to be an interesting development not only in the specific context of those respective IGOs and former ISOs, but also for the broader development of international space law, if only because the conflicts between commercial interests and general public interests are now more likely to lead to proper legal disputes requiring legal solutions, as opposed to being dealt with at an international political (member state) level.

At the same time it should be reiterated that the United States, the United Kingdom and France remain the primary addressees of any question of legality of the operations of Intelsat, Inmarsat and Eutelsat respectively. Beyond the core public duties addressed in the Public Service Agreements – which essentially concern entitlements for (certain) member states of ITSO, IMSO and EUTELSAT IGO – these three states will be held accountable by third party states. Here also any such disputes would more readily be handled in proper legal terms in view of the applicable domestic regimes of authorisation and continuing supervision, including national space law.

## 3. *The Rescue Agreement and Non-State Actors*

### 3.1. *Intergovernmental Organisations and the Rescue Agreement*

Where the Outer Space Treaty specifically mentioned intergovernmental organisations yet did not acknowledge or grant them any substantive measure of independent legal personality, in elaborating Articles V and VIII thereof[76] the Rescue Agreement went one step further with a view to IGOs.

---

[76] See Preamble, Rescue Agreement; further e.g. Carl Q. Christol, *The Modern International Law of Outer Space* (New York: Pergamon Press, 1982): 168–70; Roy S.K. Lee, "Assistance to and Return of Astronauts and Space Objects," in *Manual on Space Law*, ed. Nandasiri Jasentuliyana and Roy S.K. Lee, Vol. I (Dobbs Ferry: Oceana Publications, Inc.; Alphen aan den Rijn: Sijthoff & Noordhoff, 1979): 57; Paul G. Dembling and Daniel M. Arons, "The Treaty on Rescue and Return of Astronauts and Space Objects," *William and Mary Law Review* 9 (1968): 631 ff.

It provided namely that "for the purposes of this Agreement, the term 'launching authority' shall refer to ..., where an international intergovernmental organisation is responsible for launching, that organisation," if three further conditions would be fulfilled:[77] (1) the majority of the IGO's member states being parties to the Outer Space Treaty; (2) the majority of the IGO's member states being parties themselves to the Rescue Agreement; and (3) the IGO issuing a formal declaration of "its acceptance of the rights and obligations provided for in this Agreement."[78] Since the concept of the 'launching authority' pointed to the entity principally enjoying the rights proffered by the Agreement,[79] this clause essentially allowed ISOs to become, to a considerable extent, equated to states for the purpose of the Agreement's regime.

Neither INTELSAT, nor INMARSAT, nor EUTELSAT however have issued such a declaration;[80] in any case where these ISOs would have been interested in recovering a satellite they would have had to call upon one or more of their member states, or through the contract with the launching authority upon such authority, to ensure these interests were being recognised and respected. This situation has persisted with ITSO, IMSO and EUTELSAT IGO after the privatisation of the operators was finalised, although the lack of direct involvement of these IGOs with the satellites and satellite operations themselves makes this a rather theoretical point.[81] It would be hard to derive from the roles of ITSO, IMSO and EUTELSAT IGO a status of 'launching authority,' which after all is defined as being "responsible for launching" of the satellite at issue.[82] As following from earlier analysis, ITSO, IMSO and EUTELSAT IGO now act in a purely reactive mode to decisions on launching and operating satellites that are made by Intelsat, Inmarsat and Eutelsat – and then only to the limited extent of ascertaining their compliance with certain international public duties.

Finally it should be noted that the equation of IGOs to states for the purpose of the Rescue Agreement is limited to the substantial obligations

---

77 Art. 6, Rescue Agreement. See further e.g. Dembling and Arons, "The Treaty on Rescue and Return of Astronauts and Space Objects," 658–9.

78 Art. 6, Rescue Agreement.

79 Cf. Artt. 1–5, Rescue Agreement.

80 Cf. available: http://www.unoosa.org/oosatdb/showTreatySignatures.do; under 'Treaty': 'Rescue Agreement.' Last accessed 20 June 2012.

81 See also analysis *supra*, para. 2.1.2.

82 Art. 6, Rescue Agreement.

related to (rescue and) recovery of astronauts and spacecraft; the concept of the 'launching authority' does not play a role in the context of any procedural issues such as the possibilities to amend the Rescue Agreement.[83]

### 3.2. *Private Operators and the Rescue Agreement*

Contrary to the Outer Space Treaty, the Rescue Agreement does not make any reference to private parties, not even under a broad concept of 'non-governmental entities.' In addition to the general lack of perception that private enterprise might be interested in space activities and the political resistance of the former Soviet Union to legal recognition of private space activities, the absence of any reference to private parties was also due to the focus of the Agreement on manned spaceflight and assistance to astronauts, where such participation was expected least of all.[84]

For private operators, this means that relevant legal rights and obligations are completely subsumed under those of the contracting state concerned as 'launching authority,' consistent with the general approach on this set forth in particular in Article VI of the Outer Space Treaty.[85] In addition, the Rescue Agreement has been argued above to essentially constitute an elaboration of two specific Articles of the Outer Space Treaty.

As a consequence, on the side of claiming rights under the Agreement, the state(s) qualifying as the launching authority(ies) for a telecommunication satellite would have to take up the cause of the private operator in case the latter is seeking, for example, for return and recovery of its satellite. For Intelsat, at least the United States would be such a 'launching authority' and 'contracting party,' for Inmarsat the United Kingdom and for Eutelsat France.

In addition, however, possibly one or more other states might also qualify as 'launching authority.' Take the example where Intelsat would choose to launch one of its satellites from Bajkonour, a Russian facility located on

[83] Cf. Art. 8, Rescue Agreement; the procedural clauses of the Agreement are generally found in Artt. 7–10.

[84] It may be noted that four of the five substantive Articles of the Rescue Agreement deal with astronauts, only the fifth one with space objects.

[85] As Art. VI, Outer Space Treaty, *inter alia* equates, for the purpose of international responsibility and international space law generally, activities of non-governmental entities to those of the state itself, this *vice versa* justifies that state in taking up the cause of such entities in the international arena.

the territory of Kazakhstan. The concept of the 'launching authority' has not been defined by the Rescue Agreement beyond the state or IGO "responsible for launching."[86] Noting however that most experts tend to blend the notion of 'launching authority' with the later one of the 'launching state,' which *has* been defined by the Liability and Registration Conventions in greater detail,[87] any launch from Bajkonour would make both Russia and Kazakhstan launching states and therefore might also qualify them as launching authorities for the purpose of the Rescue Agreement.

The only problem here would be that the Rescue Agreement throughout refers to the concept of 'launching authority' explicitly in the singular. With, in the example given, the United States, Russia and Kazakhstan effectively all sharing the 'responsibility' for the launch, under a strict interpretation the question would arise which state would be the 'most responsible' one so as to most logically qualify as the launching authority.[88] Or should the explicit singularity of the 'launching authority' be considered effectively overridden, outdated by later developments regarding the possible plurality of the 'launching state' under the Liability and Registration Conventions?[89]

On the other hand, for those private operators that would find themselves confronted with complaints emanating from obligations following from the Rescue Agreement, the launching authority would be the relevant contracting party to the Agreement to defend against legal claims, although this would be largely theoretical. Private operators would not likely become involved in scenarios where they would be able to play a substantial role in 'rescuing' or returning a satellite. At best, it could be imagined that a satellite operator would be the first to receive or discover "information ... that

---

86 Art. 6, Rescue Agreement. See further Bin Cheng, *Studies in International Space Law* (Oxford: Clarendon Press, 1997): 279–80; Christol, *The Modern International Law of Outer Space*, 181.

87 See Art. I(c), Liability Convention, Art. I(a), Registration Convention; in both cases, the 'launching state' is defined as "(i) A State which launches or procures the launching of a space object; (ii) A State from whose territory or facility a space object is launched." Further *infra*, para. 4.2.1.

88 Cf. also a similar discussion at Wassenbergh, 22–31.

89 Similar questions were addressed with respect to the concept of the 'appropriate state' in Art. VI, Outer Space Treaty, *vis-à-vis* the possible multitude of states internationally responsible under the same Art.; see e.g. the author's *Private Enterprise and Public Interest in the European 'Spacescape,'* 18–22.

the personnel of a spacecraft have alighted on the high seas or in any other place not under the jurisdiction of any State," triggering the relevant obligations for the contracting party.[90]

In such cases, it would seem to be exclusively the United States in respect of Intelsat, the United Kingdom in respect of Inmarsat and France in respect of Eutelsat which would be held to answer such claims, although it could not be excluded that, for example, the operation of a particular Inmarsat satellite conducted from US territory might trigger US responsibility in addition to that of the United Kingdom. The concept of 'national activities' determines the scope of the international responsibility addressed by Article VI of the Outer Space Treaty, and such 'national activities' could very well encompass, next to activities conducted by nationals of the state concerned (the United Kingdom in the case of Inmarsat), activities conducted from the territory of another state concerned (in this example the United States).[91]

### 3.3. *The 'Boundary' Issue Under the Rescue Agreement*

In the absence of any reference to private operators or non-governmental entities and with 'international intergovernmental organisations' for the purposes of the Agreement defined in no greater detail than precisely by the reference to 'intergovernmental,' the crucial dividing line between IGO/ISO and private operator remains the same as that discussed with respect to the Outer Space Treaty.[92] INTELSAT, ITSO, INMARSAT, IMSO, EUTELSAT and EUTELSAT IGO thus all respectively qualify as IGOs for the purpose of the Rescue Agreement, whereas Intelsat, Inmarsat and Eutelsat clearly do not.

The main result of the privatisation process in this context then is that the actual operators of the three satellite systems since being privatised can no longer claim rights of recovery and return directly even in theory, as dependent upon a relevant declaration under Article 6, but now in any case would require their respective launching authority (or launching authorities) to do so.

---

90 Art. 3, Rescue Agreement.

91 See the discussion *supra*, para. 2.2.2.

92 Again, such an argument is further substantiated by the general perception of the Rescue Agreement as an elaboration of the Outer Space Treaty with regard to the few clauses of the latter dealing with astronauts and space objects.

In addition, they would not be held directly responsible under international space law for any possible violations of the relevant obligations, in this case referring to issues of rescue, recovery and return. However, as discussed above, when it comes to possible violations of any rules of space law under the Outer Space Treaty, in the case of an ISO there would always be a residual responsibility for those states members of the ISO, so that the difference in that respect would not be very substantial.

## 4. *The Liability Convention and Non-State Actors*

### 4.1. *Intergovernmental Organisations and the Liability Convention*

The Liability Convention is generally perceived to constitute an elaboration of Article VII of the Outer Space Treaty, establishing a more detailed set of liability rules based upon the fundamental principle of liability created by it.[93] The Convention specifically refers to "international intergovernmental organisations" in just two of its clauses: one on the specific issue of definition of damage compensable under the Convention,[94] the other having a fundamental effect on the scope of the Convention *ratione materiae* rather akin to the formula of Article 6 of the Rescue Agreement.

#### 4.1.1. *The secondary Status of IGOs Under the Convention*

As to the latter, Article XXII(1) of the Liability Convention provides: "In this Convention, with the exception of Articles XXIV to XXVII, references to States shall be deemed to apply to any international intergovernmental organisation which conducts space activities if the organisation declares its acceptance of the rights and obligations provided for in this Convention and if a majority of the States members of the organisation are States Parties to this Convention and to the Treaty on Principles Governing the

93 See e.g. Carl Q. Christol, "International Liability for Damage Caused by Space Objects," *American Journal of International Law* 74 (1980): esp. 354–6; Zhukov and Kolosov, 101; Christol, *The Modern International Law of Outer Space*, 60, 90–1; Bin Cheng, "International Liability for Damage caused by Space Objects," in *Manual on Space Law*, ed. Nandasiri Jasentuliyana and Roy S.K. Lee, Vol. I (Dobbs Ferry: Oceana Publications, Inc.; Alphen aan den Rijn: Sijthoff & Noordhoff, 1979): 87, 97–8.

94 Art. I(a), Liability Convention, defines damage as including "loss of or damage to property of States or of persons, natural or juridical, or property of international intergovernmental organisations."

Activities of States in the Exploration and Use of Outer Space, including the Moon and Other Celestial Bodies." Member states of an IGO, furthermore, are charged specifically with trying to get that IGO to make such a declaration.[95]

The Convention thus requires the same three conditions to be fulfilled as in the case of the Rescue Agreement. Further, like the Rescue Agreement, the 'partisanship' of an IGO in its own right is limited to the substantive rights and obligations, where Articles XIV through XX deal with the Claims Commission as the dispute settlement mechanism offered by the Convention to states parties. Article XXI deals with the possible need to render assistance in case of large scale dangers to human life and Articles XXIII through XXVIII deal with procedural issues and formalities.[96]

More clearly than in the Rescue Agreement, however, the legal personality potentially granted to ISOs is of a secondary nature. When the ISO in its quality of 'launching state'[97] has become liable for damage caused by its space object, any claim for such damage should be addressed firstly to the organisation itself, yet in a second instance, "where the organisation has not paid, within a period of six months, any sum agreed or determined to be due as compensation for such damage, … the claimant State [may] invoke the liability of the members which are States Parties to this Convention for the payment of that sum."[98] Six months, however, is not a terribly long period for an ISO to deal with an international liability claim under the Liability Convention.

In other words: behind the 'surface' liability of an ISO there would always lie a 'baseline' liability of the individual member states; such states can never use the ISO as a shield against any obligation to see the victim duly compensated and would be instead be incentivised to arrange for a proper handling of such claims in the context of the IGO prior to such event.[99]

---

95 See Art. XXII(2), Liability Convention.

96 E.g., Art. XXIII, Liability Convention, determines that the Convention leaves untouched any other relevant international agreement between states, and Art. XXV provides for the right to propose amendments to the Convention.

97 See Art. I(c), Liability Convention, for the four alternative criteria to being a "launching State" for the purpose of the Convention; also *infra*, para. 4.2.1.

98 Art. XXII(3), sub (b), Liability Convention.

99 Cf. the case of the European Space Agency (ESA) (established by means of the Convention for the Establishment of a European Space Agency, entered into force 30 Oct. 1980, 14 I.L.M. 864 (1975)), which declared its acceptance of rights and obligations under the Liability Convention as per a Declaration of 23 September 1976 (Space Law – Basic Legal

Such arrangements would have to deal most importantly with questions (1) regarding contributions to any liability claim to be paid by an ISO by those of its member states *not* parties to the Liability Convention (and hence to that extent not under a self-evident obligation to carry their share of the ISO's liability), and (2) whether the proportionality criterion to be used for such cases should refer to the overall respective contributions made to the ISO, the respective contributions made to the particular satellite, a general calculation of GNP or any combination of those and/or other criteria. Thus, if the ISO is unable to pay the claim directly, the victim could still seek compensation from the member state.

On the other hand, if an ISO would become the victim of damage compensable under the Convention, its legal status is even more secondary, as "any claim, pursuant to the provisions of this Convention, for compensation in respect of damage caused to an organisation which has made a declaration in accordance with paragraph 1 of this article shall be presented by a State member of the organisation which is a State Party to this Convention."[100]

Clearly, an ISO is not an autonomous entity whose intergovernmental composition is irrelevant, presenting a public version of a 'corporate veil' hiding the individual interests of the 'shareholders.'[101] As seen before in the context of Articles VI and XIII of the Outer Space Treaty, the ISO in such cases is considered in essence to be a vehicle for a number of states to act collectively, stimulating states to use it also as a vehicle for dealing with joint liabilities for the activities of the ISO.

---

Documents, A.III.2, at 1), and set out to elaborate the internal consequences by means of a Resolution on the Agency's Legal Liability, ESA/C/XXII/Res. 3, adopted Paris, 13 Dec. 1977 (International Organisations and Space Law, 1999): 35. See further Bohlmann and Süss, "The status of international intergovernmental organisations under the UN Outer Space Treaty System," 9.

100 Art. XXII(4), Liability Convention.

101 Allusion is made here to the famous Case Concerning the Barcelona Traction Light and Power Company, Limited (Second Phase) (Belgium v. Spain), International Court of Justice, Febr. 5, 1970, I.C.J. Rep. 1970, 4; where the ICJ used the concept of the 'corporate veil' created by the establishment of the Barcelona Traction Light and Power Company, a 'veil' not to be 'pierced,' as the key argument for not allowing Belgium to exercise diplomatic protection on behalf of the Belgian shareholders of the Company (who constituted the overwhelming majority thereof). See further e.g. Bohlmann and Süss, "The status of international intergovernmental organisations under the UN Outer Space Treaty System," 9; Shaw, 816–8; Malanczuk, *Akehurst's Modern Introduction to International Law*, 266–7.

### 4.1.2. *The Special Case of EUTELSAT IGO*

Among the three original ISOs which are the object of current analysis, EUTELSAT is the only one to have actually declared its acceptance of rights and obligations under the Liability Convention by means of a relevant declaration.[102] While that declaration may persist also with respect to the EUTELSAT IGO which took the place of the old 'EUTELSAT,'[103] it remains to be seen what this means in light of the fact that any space operations of note – the kind of activities potentially triggering application of the Liability Convention – have been transferred to Eutelsat in the course of the privatisation process.

On the 'defendant-side,' with a view to the definition of the "launching State," applicability of the Convention could only become an issue to the extent that EUTELSAT IGO would itself 'procure' the launch of another satellite on behalf of Eutelsat.[104] Such a situation, however, would now be highly unlikely, as the main legislative efforts driving privatisation of the ISOs called for operational independence of the private operator from the supervising public entity. For Europe, the 1994 EU Satellite Directive required full-fledged privatisation of the ISOs *inter alia* through deletion of any special or exclusive rights, state aids of any sorts as well as other benefits IGOs could derive from such a status.[105] Likewise, for the United States

---

102 As per 30 November 1987. See United Nations treaties and principles on outer space and related General Assembly resolutions, Addendum, Status of international agreements relating to activities in outer space as at 1 January 2009; ST/SPACE/11/Rev.2/Add.2, at 16; also e.g. Nandasiri Jasentuliyana, "The Future of International Telecommunications Law," in *Legal Visions of the 21st Century: Essays in Honour of Judge Christopher Weeramantry*, ed. Anthony Anghie and Garry Sturgess (The Hague: Kluwer Law International, 1998): 399 at n. 26.

103 Nothing can be found in the EUTELSAT Convention as amended to suggest that, in transferring all operational activities and related assets from EUTELSAT to Eutelsat, the remaining EUTELSAT IGO has taken or would take steps to disavow the declaration made by EUTELSAT in 1987.

104 Note that the definition of 'procurement' is also far from universally agreed upon; cf. Gerhard, "The State of the Art and Recent Trends in the Development of National Space Law," 67–8; Michael Chatzipanagiotis, "Registration of Space Objects and Transfer of Ownership in Orbit," *Zeitschrift für Luft- und Weltraumrecht* 56 (2007): 235.

105 Commission Directive amending Directive 88/301/EEC and Directive 90/388/EEC in particular with regard to satellite communications, 94/46/EC, Oct. 13, 1994, OJ L 268/15 (1994); see Artt. 2, 3, applying Commission Directive on the competition in the markets of telecommunications services, 90/388/EEC, June 28, 1990, OJ L 192/10 (1990), to the satellite sector. See further e.g. Silvia Ospina, "International Satellite Service Providers,"

the 2000 ORBIT Act[106] required privatisation to be comprehensive in order to allow fair competition of the privatised former ISOs with private operators that did *not* evolve from ISOs, such as PanAmSat, which had lobbied vehemently with the US government to get it to ensure a level playing field in those respects.[107]

On the 'claimant-side,' EUTELSAT IGO by its declaration would still qualify as an IGO entitled under Article XXII of the Convention to the rights and duties of states under the Convention, but, as discussed, for assertion of a claim for compensation of damage suffered, it would depend upon a member state. The sole difference being that, were EUTELSAT IGO to be treated as the entity itself suffering damage (namely 'through' the private operator it is monitoring following the privatisation of the operations, as if by proxy), any EUTELSAT IGO member state could be called upon to assert a claim on behalf of Eutelsat. However, if such damage could not be legally constructed so as to give rise to implementation of the EUTELSAT declaration, only France (as the state of nationality of Eutelsat) would be entitled to put forward such a claim.

### 4.2. *Private Operators and the Liability Convention*

Like the two other space treaties discussed so far, the Liability Convention does not specifically refer to private enterprise, although it does make reference to 'juridical persons' a few times in the context of defining compensable damage under the Liability Convention.[108]

---

in *Proceedings of the Project 2001 Workshop on Telecommunication* (Cologne: Institute of Air and Space Law, 2000): 155–6; Herbert Ungerer, "Transformation of ISOs: European Perspective," *Outer Space Z News* 2-2 (1999): 13–6.

106 Open-market Reorganisation for the Betterment of International Telecommunications Act (hereinafter ORBIT Act), Public Law 106–180, 106th Congress, Mar. 17, 2000; see Secc. 2, 601–602, also Secc. 622–625, 641–643, 661.

107 See Patrick A. Salin, "An Illustration Of The Privatisation Process Of Outer Space," *Zeitschrift für Luft- und Weltraumrecht* 50 (2001): 217–36; further Bohlmann, Schrogl and Zilioli, 218–9; Patrick A. Salin, "Impact of Recent US Legislation and Regulations on International Satellite Communication Regulations," *Zeitschrift für Luft- und Weltraumrecht* 48 (1999): 52–5; Salin, *Satellite Communications Regulations in the Early 21st Century*, 202 ff., 487–9.

108 Thus, Art. I(a), Liability Convention, makes reference to "loss of or damage to property of States or of *persons*, natural or *juridical*," emphasis added; whereas Art. VIII refers to the *jus standi* of a state for its juridical persons. In addition, Art. VI refers to the concept of 'juridical persons' in the context of exoneration from absolute liability.

### 4.2.1. *The Proper Place of Private Operators Under the Convention*

On the 'liability-side,' as a consequence of the four criteria to become a liable entity defined as "launching State" (every "State which launches or procures the launching of a space object … [or] from whose territory or facility a space object is launched" qualifies as a "launching State," and hence is liable for damage caused by the space object concerned under the Convention)[109] no measure of private involvement in the space object concerned, whether it concerns its manufacture, its launch, its ownership, its operation or its usage, is relevant at the international level. Regardless of such private involvement, a particular state will be held liable under the Liability Convention once it qualifies as a launching state, and whether and to what extent it would be able to have recourse to any private entity involved is not a matter of the Convention itself, or for general international law for that matter.

The Liability Convention, in contrast to the Rescue Agreement and its concept of 'launching authority' as discussed before, allows for multiple states to qualify as a 'launching state,' and specifically establishes joint and several liability in cases where more than one state qualifies as such.[110] Thus, if Intelsat were to launch a satellite from Bajkonour, the United States as well as Russia and Kazakhstan would qualify as launching states. The resulting joint and several liability is a matter for the states concerned to deal with. For example, Russia and Kazakhstan have concluded an ongoing agreement that any liability claims addressed at Kazakhstan as a consequence of Bajkonour-launches will be reimbursed by Russia.[111] This bilateral inter-party arrangement, however, clearly cannot derogate from the liability as such of Kazakhstan under the Liability Convention.

The issue of reimbursement by the private operator, the *de facto* 'causator' of the damage (as well as the establishment of other means of legal control over, and involvement with, such private participation) as a consequence of the definition of the liable entity (entities) through the concept of the 'launching state' becomes a matter of domestic action. Notably, this

---

109 Art. I(c), Liability Convention; cf. further Artt. II–V especially.

110 Cf. Artt. IV, V, Liability Convention.

111 Treaty between the Government of Russia and the Government of the Republic of Kazakhstan on the Leasing of the Baikonour-Complex, done Dec. 10, 1994; see Mahulena Hosková, "The 1994 Baikonour Agreements in Operation," in *Proceedings of the Forty-Second Colloquium on the Law of Outer Space* (Reston: AIAA, 2000): 265–8.

would take the form of the drafting of national space legislation or national licenses providing for a reimbursement obligation, and – if considered requisite – attendant obligations of financial security, insurance, technical and operational expertise, and appropriate other measures to limit the opportunities that damage would actually be caused.

And indeed, in a number of instances individual states have taken care to oblige satellite operators (as well as other private companies active in outer space so as to run the risk of causing damage compensable under the Liability Convention) (1) to take out a license before being allowed to commence or continue their activities; (2) to accept in such a license the obligation to reimburse the licensing state for claims the latter might have to pay out under the Liability Convention, either without or up to a limit; and (3) to insure, at least up to a certain amount, against such a reimbursement obligation.[112]

The Liability Convention, however, by its approach to handling liability, gives rise to a peculiar situation here: essentially it is not a space activity causing damage in a particular case which leads to the liable entity or entities, but the space object involved in that activity, where the launch then points to the liable entity or entities.[113] That complete focus on the launch, however, and on the state(s) crucially involved therein as per the fourfold criterion, essentially means 'once a launching state, always a liable state.'[114] In turn, in domestic legal mechanisms devised to handle such liability in cases of private involvement, the focus will be largely on such involvement in the launch, as opposed to, for instance, the satellite operations themselves.

---

112 For a very extensive discussion of such general issues, see e.g. Pamela L. Meredith and George S. Robinson, *Space Law – A Case Study for the Practitioner* (Dordrecht, Boston: Martinus Nijhoff, 1992): esp. 71–155, 292 ff.; Valérie Kayser, *Launching Space Objects: Issues of Liability and Future Prospects* (Dordrecht: Kluwer Academic, 2001): e.g. 12 ff.

113 See also Chatzipanagiotis, e.g. 237–8.

114 The Liability Convention does not take any other state into consideration for apportionment of liability than the state(s) involved in the launch as 'launching states,' and neither Liability Convention nor Registration Convention have allowed for qualification as launching / liable state after the launch, e.g. by in-orbit take-over. Even more pointedly, the Registration Convention does not even formally allow for re-registration. In other words: there is no formal manner to get rid of a status of 'launching state' once acquired. See further on these issues Chatzipanagiotis, 229 ff.; Gerhard, "The State of the Art and Recent Trends in the Development of National Space Law," 66.

### 4.2.2. *Private Operators (such as Intelsat) and US National Space Law*

This is most clear in the US case, relevant to Intelsat, as the United States has different licensing regimes for private involvement in launching (and other forms of space transportation), satellite communications[115] and satellite remote sensing[116] respectively, whereas the regime providing for liability-reimbursement and related insurance obligations is incorporated exclusively in the first-mentioned regime.

Here, the Commercial Space Launch Act Amendments of 1988, now codified in their latest rendition as part of the United States Code,[117] provided that anyone licensed under the Act "shall obtain liability insurance or demonstrate financial responsibility in amounts to compensate for the maximum probable loss from claims by ... a third party for death, bodily injury, or property damage or loss resulting from an activity carried out under the license."[118] That insurance is to cover *inter alia* the US government against such claims as may arise under the Liability Convention.[119] The 'maximum probable loss' referred to in this key clause is to be determined by the Secretary of Transportation (read the Office of the Associate Administrator for Commercial Space Transportation as mandated by the former), but any possible amount to be derogated under this concept would be capped by either "the maximum liability insurance available on the world market at reasonable cost" or US$ 500,000,000, whichever is the lower.[120]

Most importantly, however, the particular licensing obligation entailing such further responsibilities applies only to anyone wishing "*to launch a*

---

[115] As essentially based on the 1934 Communications Act; see further *supra*, para. 2.2.3.

[116] First as per the Land Remote Sensing Commercialization Act, Public Law 98–365, 98th Congress, H.R. 5155, July 17, 1984, 98 Stat. 451; then as per the Land Remote Sensing Policy Act, Public Law 102–555, 102nd Congress, H.R. 6133, Oct. 28, 1992, 15 U.S.C. 5601, 106 Stat. 4163.

[117] Commercial Space Transportation – Commercial Space Launch Activities, 51 U.S.C. Chapter 509. See further Marboe and Hafner, 40–2; Petra A. Vorwig, "Regulation of Private Launch Services in the United States," in *National Regulation of Space Activities*, ed. Ram S. Jakhu (Dordrecht: Springer, 2010): 405–19; Kayser, 90–134.

[118] Sec. 50912(a) (1), Commercial Space Transportation – Commercial Space Launch Activities.

[119] Cf. Sec. 50912(a) (4) (A), Commercial Space Transportation – Commercial Space Launch Activities. The insurance is also to cover the tort (third-party) liability that arises under domestic US law *vis-à-vis* any applicant, suing in a private capacity in US courts.

[120] Sec. 50912(a) (2) & (3), Commercial Space Transportation – Commercial Space Launch Activities.

*launch vehicle or to operate a launch site* ... in the United States," as well as any "citizen of the United States ... [wishing] *to launch a launch vehicle or to operate a launch site* ... outside the United States" – not, for example, to someone only operating the satellite launched and/or having ordered such launch.[121]

Any private satellite operator, even if a US company in legal terms (like Intelsat) and operating its satellites from US soil (as Intelsat at least partially does), would at best come in touch with these rules indirectly. This would occur namely only to the extent it would have its satellites launched by an operator that is subject to the above licensing requirements, and may wish to derogate, as part of the launch contract, (part of) those liabilities to the satellite operator. Most likely, relevant provisions would focus on damage caused as a consequence of negligence or gross negligence on the part of the launch provider alternatively satellite operator, noting that the baseline standard of practice in such contracts is a cross-waiver of liability, or on cases where the damage was caused unequivocally after the launch phase had ended and the satellite was on its own. Ultimately, however, whatever provisions end up in the contract are a matter of negotiation between the two contractual partners, not a matter of international or (generally speaking) of national space law.

### 4.2.3. *Private Operators (such as Inmarsat) and UK National Space Law*

In contrast to the United States, the United Kingdom and France have single 'framework' acts dealing *ratione materiae* with the comprehensive range of space activities, including satellite operations. Inmarsat, a body "incorporated under the law of any part of the United Kingdom," not being a launch provider itself has to procure the launches of its satellites by launch service providers (whether these are governmental or private), yet under the UK Outer Space Act it does indeed require a license for each such activity.[122]

The Outer Space Act then also imposes upon the licensee the obligation to "indemnify Her Majesty's government in the United Kingdom against any claims brought against the government in respect of damage or loss

121 Sec. 50904(a) (1) & (2), Commercial Space Transportation – Commercial Space Launch Activities; emphasis added. Cf. also sub (3) & (4).

122 Sec. 2(1), Outer Space Act, in conjunction with Sec. 1(a), requiring any British company which is "procuring the launch of a space object" to obtain a license.

arising out of activities carried on by him to which this Act applies."[123] As for insurance, it is left to the licensing authorities to impose such an obligation as well as any limit to it.[124] At the policy level it has become standard practice to impose such an obligation, with moreover a standard cap on third-party liability insurance being applied of (currently) some € 60,000,000, some US$ 80,000,000 at today's rates.[125]

In view of the aforementioned focus of the international liability regime as per the Liability Convention on the launch for the purposes of determining liability, it may appear unlikely that the UK authorities would impose such an insurance requirement for a satellite operator (as opposed to a launch operator) without further ado.

However, the reference to 'procurement' of a launch as itself requiring a license under the Act may signify that the procurement of a launch by a private UK company such as Inmarsat should be read as procurement of that launch by the United Kingdom itself, making that state liable under the Liability Convention in case damage would be caused by the satellite thus launched. To be sure, Article I(c) sub (i) of the Liability Convention speaks of "a *State* which ... procures the launching of a space object;" yet many experts would hold the reference to 'a state' to include any private company subsumed under that state, in line with the general equation by Article VI of the Outer Space Treaty of private activities with governmental activities for the purpose of international space law.[126] Consequently, Inmarsat may indeed be required under its license to take out insurance,

---

123 Sec. 10(1), Outer Space Act; subsec. (2) only provides for two specific exceptions to this special rule which are not relevant here.

124 See Sec. 5(2) (f), Outer Space Act.

125 Cf. the UK Space Agency's statement on the issue, available: http://www.bis.gov.uk/ukspaceagency/what-we-do/space-and-the-growth-agenda/uk-capabilities-for-overseas-markets/the-outer-space-act-1986. Earlier policy statements referred to an amount of £ 100,000,000, almost double the present applicable amount; see the 2010 Revised Guidance For Applicants Outer Space Act 1986, 2, available: http://www.bis.gov.uk/assets/bispartners/ukspaceagency/docs/osa/guiforapp2010.pdf.

126 Emphasis added. See for this discussion e.g. Armel Kerrest de Rozavel, "Launching Spacecraft from the Sea and the Outer Space Treaty: The Sea Launch Project," *Air & Space Law* XXIII (1998): 18–21; the author's "Sovereignty Versus Space – Public Law and Private Launch in the Asian Context," *Singapore of International & Comparative Law* 5 (2001): 38–42; Venturini, 59–60; Karl-Heinz Böckstiegel, "The Term 'Launching State' in International Space Law," in *Proceedings of the Thirty-Seventh Colloquium on the Law of Outer Space* (Reston: AIAA, 1995): 80–3.

essentially because the UK authorities do not exclude the possibility that their liability may be called upon internationally so as to lead to the obligation to compensate damage caused by an Inmarsat satellite.

#### 4.2.4. *Private Operators (such as Eutelsat) and French National Space Law*

Similarly in France, Eutelsat would be subject to a licensing obligation, as there is no doubt that the company qualifies as a "juridical person whose headquarters are located in France ... intending to procure the launching of a space object" in addition to being a "French operator intending to command such an object during its journey into space."[127] 'Operator' is defined as "any natural or juridical person carrying out a space operation under its responsibility and independently," whereas the term 'space operation' includes "any activity consisting ... of ensuring the commanding of a space object during its journey in outer space."[128]

Under Article 4 of the Law on space activities Eutelsat has to be pronounced financially, professionally and technically healthy by the French authorities in order to have a request for a license positively considered. More importantly, it has to "have and maintain, as long as it can be held liable pursuant to Article 13 and for the amount set out in Articles 16 and 17, insurance or another financial guarantee," which "must cover the risk of having to compensate for the damages that could be caused to third parties" up to the relevant amount, and must cover *inter alia* "1° The Government and public bodies; 2° The European Space Agency (ESA) and its Member States; 3° The operator and the persons having taken part in the production of the space object or in the space operation" – of course each to the extent these may be faced with claims for compensation.[129] The reference to ESA and its member states under 2° takes into account in particular the potential for Arianespace launches to lead ESA and/or its member states to become liable,[130] whereas the reference under 3° covers *inter alia*

---

127 Art. 2(3), Law on space activities.

128 Art. 1(2), resp. (3), Law on space activities.

129 Art. 6(I), resp. (II) & (III), Law on space activities.

130 Arianespace launches take place (so far exclusively) from Kourou, French Guyana, which is French territory; part of the launch facilities there however belong to ESA, hence making ESA, and in a subsidiary fashion all of its member states, co-liable under Art. I(c), Liability Convention. ESA, it should be noted, has deposited a declaration in accordance with Art. XXII: the Declaration of 23 September 1976 (International Organisations and Space Law (1999): 33.

product liability at the international level, something not as such addressed by the Liability Convention. Interestingly, Article 15 allows operators from other member states of the European Union (plus those of the European Economic Area)[131] the same possibility as a French operator to avail itself of the maximum liability arrangements under a license.[132]

In view of the quoted reference to Article 13, Article 6 essentially reiterates some of the provisions of the Liability Convention regarding absolute *versus* fault liability as well as exoneration. Articles 16 and 17 provide the French authorities with the ability to limit the indemnification for damage caused during the launching phase respectively thereafter.

Finally, Article 14 ensures that the French government, if found liable, will be reimbursed by the licensee up to the maximum amount applicable under Articles 16 or 17. Interestingly, that assurance not only applies to international liability claims under the Liability Convention, but also to such claims under the Outer Space Treaty. This suggests that France at least does not wish to exclude the possibility that a third state would like to claim *not* on the basis of the Liability Convention (and therefore effectively also *not* on the basis of Article VII of the Outer Space Treaty)[133] but on the basis of Article VI.[134]

Apparently, therefore, the mere conduct of space operations by Eutelsat, without these necessarily leading to qualification of France as a 'launching state' may, if causing damage, from the French perspective *still* lead to a claim for compensation, as 'reparation' for the consequences of 'national activities in outer space' attributable under Article VI to France.

Take, for instance, the situation in which Eutelsat would buy a satellite already in orbit where it had no involvement whatsoever with the launch thereof (for example, buying a satellite from a US competitor that

131 The European Economic Area was created by the Agreement on the European Economic Area, entered into force 1 Jan. 1994, OJ L 1/3 (1994), to extend the scope of the larger part of the European Union's Internal Market regime to a few other European states, notably (as of today) Iceland, Liechtenstein and Norway.

132 In this sense, the French national space law was the first to explicitly conform to a fundamental principle of European Community law; that there is to be no discrimination between companies from one EU member state and those from another.

133 Only in cases where the third state is party to the Outer Space Treaty only but not to the Liability Convention would it make sense to use Article VII of the former as the basis for a claim instead of the comprehensive clauses of the latter.

134 See on this specific issue the author's "Liability versus Responsibility in Space Law: Misconception or Misconstruction?," 363–7.

had procured its launch from a US launch service provider, launching from a US facility on US territory).[135] Under its own law, France would impose the obligation upon Eutelsat to obtain a license for operating that satellite after hand-over, including the obligation to indemnify the French government in "the case of ... damage caused after the launching phase" up to a certain maximum.[136]

### 4.2.5. *Private Operators as Victims of Damage*

On the 'victim-side,' read 'claimant-side' of liability, the primary focus of the Liability Convention on states was logically reflected in a total absence of *jus standi* for private operators – at least under the Convention itself. Any claim for damage, even if sustained by a private operator (for example, if its satellite would be damaged by fragments of another space object), would require a state to take it up. Here, the Convention firstly provides for the state of nationality of the operator, secondly (if the aforementioned state choose not to assert a claim) the state where the damage was sustained (of course presuming it a different state than the first-mentioned), and thirdly (if neither of the aforementioned states has chosen to assert a claim) the state whose "permanent residents" had sustained the damage (presuming again this to be a state different from the previous one(s)).[137]

With a view to juridical persons such as private operators of communication satellites, however, the following *caveats* regarding the application of these clauses are due. As to the second option of a state taking up the cause of a private operator, it should be noted that the most likely cases of damage being suffered by a private operator as caused by an(other) space object (and hence falling within the scope of the Liability Convention) would be to its own satellite in outer space, meaning this option essentially becomes useless: outer space is not the 'territory' of any state in the legal sense of the word.[138]

---

135 Cf. also Chatzipanagiotis, 229 ff., also n. 40 at 237.

136 Art. 17, Law on space activities; see also Artt. 2, 13 & 14.

137 See Art. VIII(1), (2) & (3), Liability Convention.

138 Cf. Art. II, Outer Space Treaty, providing "Outer space, including the Moon and other celestial bodies, is not subject to national appropriation by claim of sovereignty, by means of use or occupation, or by any other means;" further e.g. Fabio Tronchetti, "The Non-Appropriation Principle under Attack: Using Article II of the Outer Space Treaty in its Defence," in *Proceedings of the Fiftieth Colloquium on the Law of Outer Space* (Reston: AIAA, 2008): 526–36; Patricia M. Sterns and Leslie I. Tennen, "Privateering and Profiteering on the Moon and Other Celestial Bodies: Debunking the Myth of Property

Furthermore, the third option seems to focus largely on *natural* persons, which is in any event where the notion of a 'permanent resident,' entitling the state concerned to the exercise of diplomatic protection such as by taking up an international claim on its behalf, is becoming an increasingly accepted concept.[139] Even if it were to be applied here to private companies as well, the notion of 'permanent residence' in the context of a company logically speaking would point at its main seat of business operations, which is its headquarters (unless a ground station for satellite control as such would be deemed to qualify) – which in turn, under general public international law, co-determines the nationality of the company[140] (together with the law of incorporation), bringing analysis back to the first option.

In short: it would essentially be the state of nationality which is to take up the cause of a private satellite operator in the context of an international liability dispute under the Liability Convention. Questions may then arise to what extent domestic provisions, either by national law or by applicable licenses, deal with the rights and interests of private operators to see their cause being properly taken up by their state of nationality, for example by a right to be consulted and/or present at relevant proceedings. The Liability Convention, being an international public law treaty, remains completely silent on those issues.

It should be noted that nothing in the Liability Convention precludes a private operator from pursuing its own claims in other forums at its disposal, notably those of the launching state.[141] There may be several valid

---

Rights in Space," in *Proceedings of the Forty-Fifth Colloquium on the Law of Outer Space* (Reston: AIAA, 2003): 58–62; Frans G. von der Dunk et al., "Surreal estate: addressing the issue of 'Immovable Property Rights on the Moon,'" *Space Policy* 20 (2004): 152–3; Ricky J. Lee and Felicity K. Eylward, "Article II of the Outer Space Treaty and Human Presence on Celestial Bodies: Prohibition of State Sovereignty, Exclusive Property Rights, or Both?," in *Proceedings of the Forty-Eighth Colloquium on the Law of Outer Space* (Reston: AIAA, 2006): 95–8.

139 See on this e.g. Shaw, 810–4; Brownlie, 406–18, incl. discussions of the Nottebohm Case, (Second Phase) (Liechtenstein v. Guatemala), International Court of Justice, Apr. 6, 1955, I.C.J. Rep. 1955, 4.

140 Cf. e.g. Case Concerning the Barcelona Traction Light and Power Company, Limited (Second Phase) (Belgium v. Spain), International Court of Justice, Febr. 5, 1970, I.C.J. Rep. 1970, 4; further Malanczuk, *Akehurst's Modern Introduction to International Law*, 266; Shaw, 815–8; Brownlie, 419–21.

141 See Art. XI(2), Liability Convention.

reasons for doing so: apart from the dependency upon a private operator's state of nationality which recourse to the Liability Convention's system requires, private claims in a court would lead to a binding judgement,[142] may cost considerably less in terms of time and funds, and may (at least in certain jurisdictions) lead to substantially larger amounts of compensation than an international decision. To that extent, the Liability Convention is merely an additional tool for relevant states to arrive at compensation for damage sustained by themselves and/or natural or juridical persons.

### 4.3. *The 'Boundary' Issue Under the Liability Convention*

The Liability Convention, like the Rescue Agreement, defines IGOs essentially in an indirect manner, through its repeated references to states members of that organisation and their fundamental role in the context of the secondary personality these IGOs can enjoy under the Convention. In other words: for the purposes of the Liability Convention the key defining factor of an ISO is again its being composed of a number of sovereign states, as this already followed from the Outer Space Treaty's relevant provisions.

Moreover, there is no proper definition of private operators, but following the Outer Space Treaty's lead clearly Intelsat, Inmarsat and Eutelsat qualify as such for the purpose of the Liability Convention, and will of necessity implicate their respective states in any case where damage would be caused by or to their operations.

Crossing the threshold from ISOs to private companies, therefore, means that henceforth Inmarsat requires a license from the UK authorities which, *inter alia*, deals with the potential for the United Kingdom to be held liable as a consequence of Inmarsat's procurement of launches, and generally speaking ensure control of the respective authorities over the technical, operational, economic, social and ecological quality of the activities.[143]

Similarly, Eutelsat would be faced with liability reimbursement arrangements as part of its license, as a consequence of the reference of the French Law on space activities to compensation possibly due under (Article VI of the) Outer Space Treaty in conjunction with its reference to licensing

142 Whereas a decision of a Claims Commission to be established under the Liability Convention is only final and binding if both states parties to the dispute have in advance so agreed; see Art. XIX(2).

143 See *supra*, para. 4.2.3.

obligations for any operations in space and (limits to) indemnification under such a license also for damage caused after the launch phase has ended.[144]

Only in the case of Intelsat, as seen, the imposition of international third-party liability arrangements through the license (which license itself was suggested by Article VI of the Outer Space Treaty and obligatory under the Commercial Space Launch Act) would be unlikely.[145] Apparently, the US authorities do *not* consider mere operations of Intelsat involving its satellites and causing damage to possibly give rise to claims for compensation as part of reparation for internationally wrongful acts involving such damage under Article VI of the Outer Space Treaty.

Whether the position on this legal matter of the United Kingdom and France or conversely that of the United States will prevail and become the generally accepted one at the end of the day will only be decided *either* when further international legal discussion (presumably in the UNCOPUOS context) would lead to an unequivocal understanding on the scope of the international responsibility and liability of Articles VI and VII of the Outer Space Treaty and of the Liability Convention *or* upon the first decision by an international court or tribunal in a concrete legal dispute on the issue.

Crossing the threshold from ISO to private operator finally also meant that, henceforth, all three operators would depend upon their respective states of nationality *only* in case they would suffer damage and would like to use the Liability Convention to get such damage compensated. If, for whatever reason, the United States, the United Kingdom or France would not consider taking up a case under the Liability Convention on behalf of Intelsat, Inmarsat or Eutelsat for damage caused to their respective satellites, the only option left to those private operators might well be to sue in a private capacity in the courts of (presumably) the (or a) launching state.[146]

---

144 See esp. Artt. 14 & 17, Law on space activities; further *supra*, para. 4.2.4.

145 See *supra*, para. 5.2.2.

146 For Intelsat, in theory also Luxembourg (Intelsat's corporate headquarters being registered there) might try to assert a claim on the international level; cf. *supra*, para. 3.2.3; in that case however the absence of a genuine link of Intelsat's operations with Luxembourg and the absence of any relevant licensing requirement imposed by Luxembourg may minimise the chance of Luxembourg successfully doing so.

## 5. *The Registration Convention and Private Operators*

### 5.1. *Intergovernmental Organisations and the Registration Convention*

The Registration Convention, tied most closely to the Liability Convention already in terms of substance – the former's regime for registration of space objects was developed to a large extent to enhance the chances of identifying the liable entity or entities for the purposes of the latter[147] – follows the same path when it comes to providing IGOs with their own status. Also the Registration Convention was designed largely to elaborate one article of the Outer Space Treaty, in this case Article VIII, providing for the general concept of registering a space object and the possibility to exercise jurisdiction over it for the registering state.[148]

Article VII(1) of the Registration Convention provides for the same three criteria for an IGO to become a *de facto* party to its regime (partisanship of the majority of its members to the Registration Convention, partisanship of the majority of its members to the Outer Space Treaty, and a formal declaration) as well as for the same limitation to such partisanship of the IGO (namely, to the Articles dealing with substantive rights and obligations only).[149]

The result of application of the Registration Convention to an IGO having made the declaration concerned – which did not apply to INTELSAT, INMARSAT or EUTELSAT, and currently does not apply to ITSO, IMSO or EUTELSAT IGO – is that such an IGO could now officially act as

147 See Chatzipanagiotis, 235–6; Marietta Benkö and Kai-Uwe Schrogl, "The UN Committee on the Peaceful Uses of Outer Space: Adoption of the Resolution on Enhancing Registration Practice and of the UNCOPUOS Space Debris Mitigation Guidelines," *Zeitschrift für Luft- und Weltraumrecht* 57 (2008): 336.

148 See e.g. Gabriel Lafferranderie, "Jurisdiction and Control of Space Objects and the Case of an International Intergovernmental Organisation (ESA)," *Zeitschrift für Luft- und Weltraumrecht* 54 (2005): 228–9; Yun Zhao, "Revisiting the 1975 Registration Convention: Time for Revision?," in *Proceedings of the United Nations/Republic of Korea Workshop on Space Law* (Vienna: UNOOSA, 2004): 127; Christol, *The Modern International Law of Outer Space*, e.g. 213–4, 218–23; Aldo A. Cocca, "Registration of Space Objects," in *Manual on Space Law*, ed. Nandasiri Jasentuliyana and Roy S.K. Lee, Vol. I (Dobbs Ferry: Oceana Publications, Inc.; Alphen aan den Rijn: Sijthoff & Noordhoff, 1979): 176.

149 See Art. VII(1), Registration Convention, referring to Artt. VIII-XII, which dealt again with such procedural issues as signature (Art. VIII(1)), ratification (Art. VIII(2)), and amendment (Art. IX).

registration entity of its own satellites (or other space objects) in conformity with its provisions. At the same time, the inherent limitations to applying the registration concept to IGOs becomes apparent in dealing with the major legal consequence of registration: the possibility to exercise jurisdiction, on a quasi-territorial basis as it were, over the space object so registered.[150] Obviously, an IGO cannot exercise jurisdiction in the sense of public international law, so the question would be who *could* do so over a space object registered by an IGO, as it were in its stead or on its behalf.[151]

Here the only clue the Registration Convention provides is that "where there are two or more launching States in respect of any such space object, they shall jointly determine which one of them shall register the object in accordance with paragraph 1 of this article, bearing in mind the provisions of article VIII of the [Outer Space] Treaty ... without prejudice to appropriate agreements concluded or to be concluded among the launching States on jurisdiction and control over the space object and over any personnel thereof."[152] This clause, obviously, is targeted at cases where two or more *states* qualify as launching states, where the main rationale is to preclude as far as possible conflicts of jurisdiction between several states.

A similar scenario however would arise where an IGO qualifies as a 'launching state' for the purpose of the Registration Convention, since in the absence of any further regulation all member states of that IGO, or at least those that are themselves parties to the Registration Convention – essentially regardless of whether in addition such states themselves would qualify as 'launching states' – might claim to exercise jurisdiction over the space object registered by 'their' IGO. So, it would be appropriate to apply the same rationale, as further corroborated by Article XIII of the Outer Space Treaty,[153] and hence essentially the same principle, in that

150 Cf. already Art. VIII, Outer Space Treaty (of which, as argued, the Registration Convention is an elaboration), which states: "a State Party to the Treaty on whose registry an object launched into outer space is carried shall retain jurisdiction and control over such object, and over any personnel thereof, while in outer space or on a celestial body."

151 See extensively Lafferranderie, "Jurisdiction and Control of Space Objects and the Case of an International Intergovernmental Organisation (ESA)," 232–9, on the case of the European Space Agency.

152 Art. II(2), Registration Convention.

153 Art. XIII, Outer Space Treaty, provides in relevant part: "Any practical questions arising in connection with activities carried on by international intergovernmental organisations in the exploration and use of outer space, including the Moon and other celestial bodies, shall be resolved by the States Parties to the Treaty either with the appropriate

"appropriate agreements" can or should be concluded among the states concerned "on jurisdiction and control over the space object and over any personnel thereof," to use the phrases of the Registration Convention.[154] Presumably, such 'appropriate agreements' would be part of the IGO's internal statutes, or being adopted under those.

Obviously, with none of the three IGOs currently under consideration either in their old or in their new manifestation having declared their acceptance of the rights and obligations under the Convention, all this has so far remained theory.

## 5.2. *Private Operators and the Registration Convention*

As for Intelsat, Inmarsat and Eutelsat, the main point of note is that the registration of satellites launched for either of them under the Registration Convention is *not* (necessarily or automatically) for respectively the United States, the United Kingdom and France to undertake; that almost exclusively depends upon what extent these three states qualify as 'launching states' for those satellites.

This would unequivocally be the case, for example, for the United States if an Intelsat satellite would be launched from a US launch site or territory, or by NASA or the US Air Force.[155] If, by contrast, that satellite would be launched by a Soyuz vehicle from Bajkonour, the only legal link between the Intelsat-owned and -operated satellite and the United States would be Intelsat's nationality as per its administrative and operational headquarters – a concept which does not figure as such in the definition of the launching state.

There would be one scenario however in this example under which the United States might still play the role of state of registry under the Registration Convention by qualifying as a 'launching state.' This depends on the extent to which the second criterion for becoming a launching state, that of 'procuring' the launch, is interpreted as including cases where a state's *national* (for instance a company with its nationality) procures the

international organisation or with one or more States members of that international organisation, which are Parties to this Treaty."

154 Art. II(2), Registration Convention.

155 Corresponding to the fourth, third and first criterion respectively for becoming a launching state under Art. I(a), Registration Convention.

launch itself, as opposed to only dealing with cases where "A *State* ... procures the launching of a space object."[156] As there is no general agreement on the international level regarding the precise scope of the notion of 'procurement,' not even amongst individual experts,[157] states have picked their own choice as to such interpretation, even if largely 'unconsciously,' by applying their respective licensing obligation and requirements regarding liability – where the concept of the 'launching state' was first developed – and registration to one or the other category of private actors.

Starting again with the example of Intelsat, the United States under the Commercial Space Launch Act does *not* require a license with the attendant reimbursement and insurance obligations as discussed[158] for operators procuring a launch on a non-US launcher from non-US soil, whereas the Communications Act's licensing system only refers to those intending to "use or operate any apparatus for the transmission of energy or communications or signals by radio" on US territory – yet does not contain any relevant liability or insurance requirements. From this, it should be deduced that the United States at least implicitly does not consider itself a 'launching state' for the purpose of either Liability or Registration Convention on the mere premise of a private operator with its *nationality* procuring the launch in question. At least, it would find a strong argument in denying any liability or obligation to register in respect of such satellite should a legal dispute on this ever come to an international court, tribunal or Claims Commission.[159]

It should be added, that in the United States there is no statutory registration of space objects which might shed further light on this issue. So far, the State Department takes care of such registration as necessary, based on information provided by other branches of the US government and referring in this context to "the official U.S. Registry of Space Objects Launched into Outer Space."[160] The Federal Aviation Administration in turn is then

156 Art. I(a), sub (i), Registration Convention; emphasis added.

157 Cf. e.g. Kayser, 34–6; Bruce A. Hurwitz, *State Liability for Outer Space Activities* (Dordrecht, Boston: Martinus Nijhoff, 1992): 22; Meredith and Robinson, 64; Bender, 280–1; Chatzipanagiotis, 235.

158 See *supra*, para. 4.2.2.

159 Most notably, of course, Artt. XIV–XX, Liability Convention, provide for such a commission to adjudicate claims regarding liability that could not be dealt with through diplomatic negotiation.

160 As per available: http://www.usspaceobjectsregistry.state.gov/.

required to assist the State Department in ensuring licensees will provide the appropriate information, which applies to "all objects placed in space by a licensed launch."[161]

Interestingly, for Inmarsat as a UK company the situation is different, as the UK Outer Space Act *does* require an operator with UK nationality procuring a launch to obtain a license, and this cannot but lead to the conclusion that the United Kingdom would consider itself a 'launching state,' and would feel itself obliged either to register the satellite at issue itself (in case the United Kingdom would be the only state qualifying as 'launching state') or to make sure *some* state does so in conjunction with other launching states, in case it is not itself the only one qualifying as such – for example if the actual launch of the satellite would be undertaken by a Soyuz from the Russian facilities at Bajkonour.[162]

The UK Outer Space Act also specifically provides that "[t]he Secretary of State shall maintain a register of space objects" and that "[t]here shall be entered in the register such particulars of such space objects as the Secretary of State considers appropriate to comply with the international obligations of the United Kingdom," thus leaving the matter essentially undecided to what categories of private satellite operators that Convention would actually apply.[163]

In the case of Eutelsat the French national space law likewise calls for a license for those operators of French nationality that wish to procure a launch and/or operate a satellite launched elsewhere. Yet, as the law on space activities also specifically refers not only to the Liability Convention but also to the Outer Space Treaty, the need for a license in the scenarios referred to may not be concluded automatically to confirm France's perception of being *liable* – to the extent the licensing obligation would be following from Article VI of the Outer Space Treaty, it would rather qualify France as a *responsible* state.

It should nevertheless be assumed that the French law does not deviate more than strictly necessary from the terms of the Liability and Registration Conventions, to both of which France is a party of long standing.[164]

---

161 14 C.F.R. § 417(19).

162 See Art. II(1), resp. (2), Registration Convention.

163 Sec. 7(1), resp. (2), Outer Space Act.

164 As for the Registration Convention, France even was the very first state to ratify, on 17 December 1975 (see 1023 U.N.T.S. (1976), at 16); it ratified the Liability Convention shortly thereafter on 31 December 1975, clearly envisaging the two treaties to form a coherent set.

Therefore, it should also be assumed that those cases for a license obligation that match the criterion of a launching state for liability purposes will be those used in the Liability Convention – and hence be identical to those of the Registration Convention.[165]

On registration itself finally, the French law simply calls for registration "in the event France has a registration obligation according to Article II" of the Registration Convention, without specifying or indicating how France itself envisions the scope of that Article with a view, for example, to private satellites and/or the operations conducted with them.[166]

### 5.3. *The 'Boundary' Issue Under the Registration Convention*

As a consequence of the above, the transition from ISO to private operator has had varying consequences as far as the Registration Convention is concerned. In the ISO context registration of satellites was rather straightforward, any member state being able to register on behalf of the ISO as being a state 'procuring' the launch of that satellite through the 'vehicle' of the ISO.[167] Following the transition, however, for Inmarsat and Eutelsat the situation became even more straightforward, with the United Kingdom considering itself to be a 'launching state' for any satellite the launch of which was procured by Inmarsat, and hence tasked to ensure proper registration, and the same applying *mutatis mutandis* to France and Eutelsat; whereas in the case of Intelsat the state or states entitled and obliged to realise registration might more readily differ from case to case and might not even necessarily include the United States.

## 6. *UNGA Resolution 37/92 and Private Operators*

### 6.1. *Intergovernmental Organisations and the UNGA Resolution*

The UN Resolution dealing with international direct television broadcasting,[168] being a Resolution of the UN General Assembly, is not a binding

165 See *supra*, §§ 3.2, 4.2.1.

166 Art. 12, Law on space activities.

167 As indicated, normally of course for practical reasons this would actually be done by the host state of the ISO at issue.

168 See further e.g. Christol, *The Modern International Law of Outer Space*, 605–10, 617 ff.; Simone Courteix, "International Legal Aspects of Television Broadcasting by Satellite,"

legal document akin to a treaty.[169] Also, the fact that it was not adopted by consensus but by a vote with most of the states of the developed world abstaining or voting against it denies the Resolution the status as being reflective of customary international law – with the exception perhaps of those states that voted in favour.[170] Thus, the substantive core of the Resolution, the perceived requirement of 'prior consent,' does not generally speaking apply to Intelsat, Inmarsat and Eutelsat, the post-ISO-privatisation operators, since neither the United States, nor the United Kingdom, nor France, the respective host states of those privatised entities, are amongst the states voting in favour of the Resolution.[171]

The UN Resolution makes reference to intergovernmental organisations exactly once, in stating that "[w]hen international direct television broadcasting by satellite is carried out by an international intergovernmental organisation, the responsibility referred to in paragraph 8 above should be borne both by that organisation *and* by the States participating in it."[172] Paragraph 8 in turn, as will be seen, simply restates the general state responsibility for space activities as already posited by Article VI of the Outer Space Treaty for the specific context concerned.

---

in *Legal Aspects of Space Commercialization*, ed. Kunihiko Tatsuzawa (Tokyo: CSP Japan, 1992): 109–10.

169 Cf. Art. 10, 11, 13 & 14, Charter of the United Nations, entered into force 24 Oct. 1945, U.S.T.S 993, 24 U.S.T. 2225; allowing the UN General Assembly to make recommendations to member states and the Security Council. See further e.g. Shaw, 1212; Cassese, 278; Wallace, 29; also Vladimir Kopal, "The Role of United Nations Declarations of Principles in the Progressive Development of Space Law," *Journal of Space Law* 16 (1988): 8.

170 It was adopted on 10 December 1982 with 107 to 13 votes, and 13 states abstaining. See David I. Fisher, *Prior Consent to International Direct Satellite Broadcasting* (Dordrecht, Boston: Martinus Nijhoff, 1990): 45–6, esp. n. 11. Other sources indicate a different vote however: *Law of Outer Space*, ed. Stephan F. von Welck and Renate Platzöder (Baden-Baden: Nomos, 1987): 606–7, record 108 in favour, 13 against, 13 abstaining; Rüdiger Wolfrum, "Direct-Broadcasting-Satellites," in *Handbuch des Weltraumrechts* [Handbook of Outer Space Law], ed. Karl-Heinz Böckstiegel (Cologne: Carl Heymanns, 1991): 405, records 108–12–13; Space Law – Basic Legal Documents, B.I.4.2, even adds up to 88–15–11 only. See further e.g. Lyall and Larsen, 48–50.

171 Both Fisher, and Von Welck and Platzöder record the United Kingdom and the United States as voting against, with France abstaining, whereas Space Law – Basic Legal Documents records all three states as voting against. Wolfrum does not indicate any specific state votes. With respect to Intelsat, its links with Luxembourg do not change the equation, as also Luxembourg belonged to the states voting against.

172 Princ. 9, G.A. Res. 37/92; emphasis added.

As the Resolution also explicitly considers the activities it deals with to be subject to the Outer Space Treaty,[173] these paragraphs simply confirm the ultimate responsibility of UN member states for compliance with the Resolution's principles in cases where intergovernmental organisations that those states are members of undertake the broadcasting activities at issue. This, of course, would only be the case to the extent those principles can be considered binding for any individual state concerned, either because it has voted in favour or because a certain obligation could be considered to reflect customary international law (or both). Further analysis might be required as regards the question whether that would actually apply to (a majority) of the member states of ITSO, IMSO and/or EUTELSAT IGO.

### 6.2. *Private Operators and UNGA Resolution 37/92*

The key provision in the Resolution regarding private operators is to be found in the aforementioned Paragraph 8, stating that "States should bear international responsibility for activities in the field of international direct television broadcasting by satellite *carried out* by them or *under their jurisdiction* and for the conformity of any such activities with the principles set forth in this document."[174]

Thus representing a specific application of the general state responsibility principle of Article VI of the Outer Space Treaty, the activities of any operator under the jurisdiction of a state should trigger the responsibility of that state just as if it concerned its own activities. The main difference with the formulation of Article VI of the Outer Space Treaty from that perspective lies in the use of the 'classic' international law-phrase "under the jurisdiction" instead of the reference to "national activities" that Article VI uses. Thus, the formulation of Paragraph 8 seems more clear in that this would normally refer to both territorial jurisdiction and jurisdiction over nationals.[175]

In the specific cases of Intelsat, Inmarsat and Eutelsat, there should be little doubt that (at least) the United States, the United Kingdom and

173 See Princ. 4, G.A. Res. 37/92.

174 Emphasis added.

175 Under general international law, jurisdiction of a state over its territory and jurisdiction over its nationals constitute the two versions of jurisdiction universally recognised and hence, as to the principles behind it, uncontested; see e.g. Wallace, 111–9; Shaw, 645–85, esp. 652 ff.

France respectively are internationally responsible for their activities with a view to the Resolution's terms. Since, however, as indicated, neither of these three states have voted in favour of the Resolution when it was enunciated in 1982, nor have they ever since given evidence of a change of legal views in this regard, the three private satellite operators currently under scrutiny would only have to comply with those principles pronounced by the UN Resolution which would reflect customary international law or would otherwise be binding upon them.[176]

### 6.3. *The 'Boundary' Issue Under UNGA Resolution 37/92*

In view of the above, the transition from ISOs to private operators operating under a limited supervision regime of revamped IGOs did not give rise to a major difference with regard to the legal obligations resting upon, or legal rights bestowed upon, the entities in question. At most it could be said at this point that the former operational ISOs would more likely have been required to operate in conformity with all the principles throughout – but essentially for political reasons: because a major portion of member states would have voted themselves in favour of the Resolution and would be willing to impose the consequences of strictly abiding by its terms as if it concerned a binding treaty, in many instances it would have been practically impossible for the ISOs to overstep those terms.

The general lack of willingness on the part of the United States, United Kingdom and France to consider the principles (or at least some of them) valid legal rules, by contrast, might have lead to the privatised operators being much less likely to be held to the principles provided by the Resolution. The final answer to that issue, however, would require an analysis of the respective licenses, where the authorities concerned would (or at least should) have taken care to somehow reflect those principles they would consider binding (as a matter of customary law) whilst omitting any of those they would not consider valid legal rules.

It may be noted finally that the various legal documents allowing the respective supervisory IGOs to force compliance of the private operators do not even explicitly refer to international law, only to the core respectively basic principles which the operators should comply with,[177] which

---

176 Cf. *supra*, § 6.1.

177 Cf. Artt. III, V; ITSO Agreement; Artt. 3, 4, IMSO Convention; Art. 2, Inmarsat PSA. Further Lyall and Larsen, 338–41, 351–2, 363–4, who at 340, n. 74, assert: "In fact the

essentially begs the question again to what extent there would be customary international law rules incorporated in the Resolution which *ipso facto* would be applicable also to the activities of the operators.

## 7. *The ITU Framework and Private Operators*

### 7.1. *The ITU Framework and Satellite Operators*

The framework regime, based upon the ITU Constitution and ITU Convention, using the ITU with its competencies and capacities as the principal instrument to deal with the most important aspects of international telecommunications, is – *of course* – by far the most important element of the international legal environment for conducting satellite communications, whether by way of hybrid public consortia such as the old INTELSAT, INMARSAT or EUTELSAT, or through privatised operators such as Intelsat, Inmarsat and Eutelsat, to the extent supervised by the new IGOs ITSO, IMSO and EUTELSAT IGO.

The present chapter does not profess to treat any part of that subject in any detail, but merely represents an effort to highlight one particular element thereof.[178] As indicated, the focus here is namely on the principled

Public Service Agreement was swept aside when Zeus Holdings bought 100 per cent of the share capital of INTELSAT [meant was Intelsat]."

178 For such more general analyses, reference might be had to Lyall and Larsen, 199–244; Lyall, *Law and Space Telecommunications*, 311–96; Ram S. Jakhu, "Legal Issues of Satellite Telecommunications, the Geostationary Orbit, and Space Debris," *Astropolitics* 5 (2007): 175–6, 180 ff.; Ram S. Jakhu & Virginia R. Serrano, "International Regulation of Radio Frequencies for Space Services," in *Proceedings of the Project 2001 Workshop on Telecommunication* (Cologne: Institute of Air and Space Law, 2000): 56 ff., 108–15; Ingo Baumann and Hans Dodel, "The ITU Filing of Satellite Systems," in *Contracting for Space*, ed. Leslie J. Smith and Ingo Baumann (Farnham: Ashgate, 2011): 367–81; Maria Buzdugan, "Recent Challenges Facing the Management of Radio Frequencies and Orbital Resources Used by Satellites," in *Proceedings of the International Institute of Space Law 2010* (Reston: AIAA, 2011): 327–33; Janet C. Thompson, "Space for Rent: The International Telecommunications Union, Space Law, and Orbit/Spectrum Leasing," *Journal of Air Law and Commerce* 62 (1996): 286 ff.; Frances Lyall, "The Role of the International Telecommunication Union," in *Outlook on Space Law over the Next 30 Years*, ed. Gabriel Lafferranderie and Daphne Crowther (Dordrecht: Kluwer Law International, 1997): 253–8; Philippe Achilleas, "Regulatory Framework for Authorising Satellite Applications: The Case of Telecommunications," in *Contracting for Space*, ed. Leslie J. Smith and Ingo Baumann (Farnham: Ashgate, 2011): 104–5.

legal differences between intergovernmental organisations and private operators, and an analysis of the international space law treaties (and the one UN Resolution dedicated to satellite communications) on this particular issue without at least a summary comparative look at the ITU framework would make little sense.

### 7.2. *Intergovernmental Organisations, Private Operators and the ITU Framework*

For the above purpose, it suffices to revisit the key mechanism for dealing with potential problems of satellite communications in the international environment, which is that of the coordination of frequencies and, indirectly, orbits.[179] The ITU framework, developed in its most original version almost a century and a half ago, for a long time has remained even more exclusively state-oriented than the UN space treaties, and this also transpires fundamentally in the aforementioned coordination process.

When states are intent on operating a satellite and on assuring the communication operations of that satellite, the process of coordination essentially consists of two steps. The first step involves the *allocation* of frequency *bands* to *categories* of communications services – or in the language of the Radio Regulations: "allocation (of a frequency band): Entry in the Table of Frequency Allocations of a given frequency band for the purpose of its use by one or more terrestrial or space radio communication services or the radio astronomy service under specified conditions. This term shall also be applied to the frequency band concerned."[180] These allocations normally

---

179 Cf. Art. 44(2), ITU Constitution, following the 1998 amendment as per Instrument amending the Constitution of the International Telecommunication Union of 22 December 1992, entered into force 1 Jan. 2000, ATS 2000, No. 8: "In using frequency bands for radio services, Member States shall bear in mind that *radio frequencies and any associated orbits, including the geostationary-satellite orbit,* are limited natural resources and that they must be used rationally, efficiently and economically, in conformity with the provisions of the Radio Regulations, so that countries or groups of countries may have equitable access to *those orbits and frequencies,* taking into account the special needs of the developing countries and the geographical situation of particular countries;" emphasis added. Further e.g. Thompson, 290–8; Ram S. Jakhu, "Legal Issues Relating to the Global Public Interest in Outer Space," *Journal of Space Law* 32 (2006): 72 ff.

180 Section 1.16, Radio Regulations. See further Art. 5, Sec. IV, for the actual Table of Frequency Allocations.

take place at the World Radio Conferences which are regularly held every two or three years, where the ITU member states collectively decide on new allocations, or changes or deletions of existing allocations, based on the perceived global and general need of specific services (satellite as well as non-satellite) to be able to use radiofrequencies for their services – and of course on their lobbying and negotiating skills.[181]

Once a frequency band is allocated to a certain category of satellite services, it is open for specific requests by individual states to be granted interference-free use of specific frequencies for uplink, downlink and, wherever applicable, inter-satellite-link purposes. Those individual requests can be forwarded to the ITU's Radio Regulations Board[182] at any moment a state has decided to go forward with a specific satellite project, the so-called 'notification.' Notification then kicks off an elaborate and often quite prolonged coordination process in order to ascertain that the new proposed satellite operation will not risk interfering with existing or previously notified requests for frequencies.[183]

The ultimate aim of the notification and coordination process is to arrive at the second step, the *allotment* of specific *frequencies* to the particular notified *service*, also in case of satellite services – or to use the language of the Radio Regulations again: "allotment (of a radio frequency or radio frequency channel): Entry of a designated frequency channel in an agreed plan, adopted by a competent conference, for use by one or more administrations for a terrestrial or space radiocommunication service in one or more identified countries or geographical areas and under specified conditions."[184]

The reference to 'administrations' as the entities entitled to the use of an allotted frequency (and therefore also the ones entitled to request allotment) is critical, since this term refers exclusively to relevant state organs: an administration is defined as "any *governmental department or service* responsible for discharging the obligations undertaken in the Constitution of the International Telecommunication Union, in the Convention of

181 In reality of course the process is much more complex, also since some allocations are exclusive, others are primary, and still others are secondary; moreover, allocations can be granted in many cases on a regional or even national basis only; cf. Art. 5, Sec. II, Radio Regulations; further e.g. Lyall, *Law and Space Telecommunications*, 347–54.

182 See Art. 14, Radio Regulations.

183 See further Lyall, *Law and Space Telecommunications*, 374–80.

184 Section 1.17, Radio Regulations.

the International Telecommunication Union and in the Administrative Regulations."[185]

Thus, neither intergovernmental organisations nor private operators have any formal independent say in either the allocation process or the allotment process, or can even request for allotment at the ITU level. Both essentially are treated equal in this respect, in that a third step, of *assignment*, becomes necessary. 'Assignment' is defined as "authorisation given *by an administration* for a radio station to use a radio frequency or radio frequency channel under specified conditions,"[186] formally speaking following the allotment of that frequency to the state whose administration is referenced.

As such, both ISOs and private satellite operators require one state firstly to go through the allotment process, and, where necessary prior to that, the allocation process, and secondly to then assign the frequencies ultimately so allotted to them in order to be able to use them under the ITU system.

The omission of intergovernmental or private operators from the key formal decision making processes within the ITU was not an oversight or neglected relic. ITU member states are explicitly held internationally responsible for telecommunication activities by entities "*authorised* by them to establish and operate telecommunications and which engage in international services or which operate stations capable of causing harmful interference to the radio services of other countries."[187] This provision literally refers to 'operating agencies,' so as to include both public and private operators, to the extent the former are not yet covered by the same obligation of Article 6(1), applicable to the member states themselves. An 'operating agency' is defined in the Annex to the ITU Constitution as "[a]ny individual, company, corporation or governmental agency," and should for the present analysis be read as including an '*international* (multi) government agency.'[188]

---

185 Section 1.2, Radio Regulations; emphasis added. See also no. 1002, Annex, ITU Constitution.

186 Section 1.18, Radio Regulations; emphasis added. In practice, of course, in many cases domestic discussions on future assignments of certain frequencies would precede (as well as guide) the relevant state's efforts to achieve allocation of those frequencies where, in case such allocation would be successfully achieved, the actual assignment might become a rather succinct formality.

187 Art. 6(2), ITU Constitution; emphasis added. Cf. also Art. VI, Outer Space Treaty, essentially providing for a similar level of international responsibility.

188 No. 1007, Annex, ITU Constitution.

While in more recent times the limited status of non-states in the ITU has started to evolve – for example, at the Kyoto Conference of 1994 by means of an amendment to the ITU Convention the possibility for non-governmental entities to participate as 'small-m' members was allowed, providing them with the right of access to all relevant information as well as consultation; and at the Minneapolis Plenipotentiary Conference of 1998 they were allowed to achieve a status of 'Sector members,' allowing for full-blown participation at the ITU sector level[189] – states remain the only full members of the organisation represented on the Council, the highest decision-making body of the ITU.[190]

### 7.3. *The 'Boundary' Issue Under the ITU Framework, as Far as Frequency Notification and Coordination is Concerned*

The only distinction between ISOs and private satellite operators in terms of obtaining a nominal guarantee of interference-free usage of usable frequencies from the ITU lies in the fact that, for IGOs, *any* member state of that IGO could propose allocation of frequency bands if needed and request allotment of specific frequencies effectively on behalf of that IGO. For a private operator there could usually only be one administration undertaking those efforts: that of the state of nationality (that is, incorporation and headquarters) of the company concerned.[191]

In practice, not even that distinction may have mattered much. In the cases of the old INTELSAT, INMARSAT and EUTELSAT, for obvious reasons of efficiency and coherence, the task of taking up the interests of these IGOs in the ITU context was delegated to the host state of the organisation – the United States, the United Kingdom and France respectively, precisely the states that post-privatisation would continue to do the same on behalf of respectively Intelsat, Inmarsat and Eutelsat. Even the inherent support including, where necessary, votes in favour of the other member states of INTELSAT, INMARSAT and EUTELSAT in the ITU decision making processes, for respective US, UK or French efforts would normally not

[189] See Art. 19, ITU Convention, as amended.

[190] See e.g. Artt. 2, 4; also Artt. 3, 8, 10, ITU Constitution.

[191] It may be reiterated here, that allocations are generally handled at the bi- or triennial ITU World Radio Conferences, that is essentially by all ITU member states collectively, whereas requests for allotment concern activities of individual states addressed on an ongoing basis to the Radio Regulations Board.

be lacking in the case of the private operators due to the existence of ITSO, IMSO and EUTELSAT IGO respectively, whose member states should still be generally interested in seeing the private operators succeed.

## 8. *Concluding Remarks*

Perhaps surprisingly, the overall conclusions regarding the privatisation process under the four space treaties, the UN Resolution and the ITU regime, are that the specific consequences of transformation from an intergovernmental organisation to a private operator summarily supervised by a scaled-down intergovernmental overseer are actually relatively marginal, and often more of a formal nature or leading to practical differences of handling the various interests involved than that there is any fundamental legal paradigm-shift to be noted.

In the final analysis, however, there is more to it than that. It is true, that even under the four space treaties, the apparent recognition of IGOs as a special type of entity, upon closer analysis, did not provide them much of a legal status and role as independent from their member states. Ultimately, states would always remain responsible and liable under the Outer Space Treaty and Liability Convention for activities of ISOs in case first-option dealings with the ISOs directly would somehow not work out to the satisfaction of the third parties invoking such responsibility or liability. Still, at least a considerable incentive was provided to prearrange internally for procedures of handling ISO liabilities – in terms of general need, similar, but in terms of structure, scope and substance quite different from the incentive for states to draft national space legislation including licensing regimes handling private operator liabilities.

For example, global public interests in satellite communications would be much more likely to be ignored or neglected in a context where a single state would determine the legal parameters for an operator's activities than in a context where a multiplicity of states would be involved in determining that framework. Conversely, the latter would have been much less likely to adapt to a rapidly changing technological, economic and social environment.

Also the seemingly independent ability to register satellites under the Registration Convention as an ISO – an option, it should be added, never used by INTELSAT, INMARSAT, EUTELSAT, ITSO, IMSO or EUTELSAT IGO – would still require the member states of such organisation to take care of the most prominent consequence of registration of a space object: the direct entitlement under Article VIII of the Outer Space Treaty and the Registration Convention to exercise jurisdiction and control over the satellite at issue. In that sense, the reality of the state-orientation of the space treaties as to a large extent still being devolved from general international public law, could not be overcome by the space treaties on their own account.

As a result, ISOs in the formal legal sense were (almost) as dependent on their member states as private space operators were on 'their' state. In both cases it is such a state, after all, which is to be held responsible and liable under key provisions of, in particular, the Outer Space Treaty and the Liability Convention.

UN Resolution 37/92 and the element of the ITU framework discussed here do not even make that much of a theoretical distinction between an intergovernmental and a private operator, hence making the difference between the pre-privatisation and post-privatisation at least in terms of structural, formal and theoretical application once more largely negligible.

The major exception worthy of note might operate more at an overarching abstract level of legal development than at the level of the operators themselves. The aforementioned 'explicitisation' of the general public duties for the privatised satellite operators by means of Public Service Agreements as noted constitutes an important measure of 'juridification' of the operations of satellite operators, the consequence of the cross-over from ISO to private operator. This has equivalents in other areas as well – for example, to comply with the requirements of the Registration Convention, the registration of satellites now requires a formalised domestic legal process instead of a political decision taken within an IGO structure, and essentially the same applies to the handling of the potential liabilities resulting from their operations under the Liability Convention in terms of now-explicit obligations to arrange appropriate insurance.

This in turn may lead to increasing diversification, as for example Eutelsat is being subjected to registration requirements even beyond the ones provided in the Registration Convention, and all three private operators are handled differently when it comes to dealing with the domestic consequences of liability under the Liability Convention. Prior to privatisation, the ISOs operated more or less directly under the *aegis* of the Liability Convention, even if INTELSAT and INMARSAT had not made a declaration under Article XXII and also the declaration of EUTELSAT did not cause it to obtain much of a status independent from its member states. Through those member states, liability applied in accordance to the terms of the Liability Convention without further ado except for those formal *jus standi* issues. Following privatisation, however, with the intermittence of domestic laws to regulate the specific liabilities of the operators, licenses could be mandated for varying activities (such as procurement in the UK case), liabilities could become limited, insurance could be made obligatory and specific obligations could be established which had never been considered under the Liability Convention (such as in the French case with product liability).

From a different angle therefore, a major result of the privatisation of the ISOs was a fragmentation and diversification of the legal framework(s) within which they had to operate. Further to this, and partly as a consequence thereof, the main differences between pre-privatisation and post-privatisation will lie in the application in practice, both administratively speaking and politically speaking, of the legal regimes discussed.

In the international arena, the private operators will have to count on their host states to a much greater extent than the ISOs used to do, for even if they still could avail themselves of support from ITSO, IMSO or EUTELSAT IGO, these IGOs obviously have only a specific and limited interest in the operators' activities, resulting in limited responsibilities and even more limited liabilities. After all, the whole idea behind the privatisation process was that, apart from some public duties which should remain guaranteed, the operators should act under as much commercial freedom as possible, meaning that the new IGOs might be less broadly and actively interested in the success of the particular operator they were supervising. Consequently, private operators have to look to their national authorities for issues of recovery of damages under the Liability Convention or for satellites to be recovered under the Rescue Agreement, which thereby becomes a matter of national law. This opens the door to diversification of applicable regimes, unless the inherent global character of the satellite communications market serves as a force promoting international harmonisation.

In the last resort, therefore, the Rubycon crossed by the privatisation of the ISOs indeed was more of a practical, economic and political character than of a strictly speaking legal character, although the legal framework did – apart from reflecting the relevant practical, economic and political paradigm changes – result in a juridification of the regimes and in considerably more precision regarding the allocation of responsibilities and liabilities. Also, as a consequence of the fact that the Rubycon was crossed, so to speak, at different fords by INTELSAT, INMARSAT and EUTELSAT, the general legal framework was seen to diversify and fragment considerably. From that perspective there is no doubt that also legally speaking the world of international satellite communications has changed profoundly following the privatisation of these three ISOs.

# Notes on Contributors

*Patricia K. McCormick* holds a Ph.D. in Mass Media/Telecommunications and a M.A. in Telecommunications from Michigan State University. She also holds a M.A. in African Studies from Howard University and a B.A. in African Studies and Geography from the University of Michigan. Her research has concerned the role of telecommunications policy and efforts to extend telecommunications services in developing countries and meet the challenge of reducing the global disparity in access to information technologies. In this regard Dr. McCormick has conducted extensive fieldwork in African and Caribbean states. Conceptually, her research is interdisciplinary in nature, viewing the telecommunications sector as integrated within a broader political and economic context at both the national and international levels. She has examined the impact of private sector participation in the International Telecommunication Union (ITU). Dr. McCormick is interested in the international diffusion of policy, specifically the privatisation and liberalisation of the telecommunications sector and the role of multilateral institutions in facilitating those policies. Recently she has examined the institutional restructuring of international satellite organisations. Currently, she is examining space debris and space surveillance. In addition to numerous conference presentations, she has published referred articles in such esteemed journals as *Telecommunications Policy*, *Gazette the International Journal for Communications Studies*, and *Science and Public Policy*. She is a member of the International Institute of Space Law (IISL) and presently serves as an Associate Professor in the Department of Communication at Wayne State University.

*Maury J. Mechanick* has more than 30 years experience in the field of international telecommunications. Since 2002, he has held the position of Counsel in the Washington, DC office of White & Case LLP, where his practice focuses on representing public and private sector clients in the satellite and telecommunications fields in transactional and regulatory matters in the United States and abroad. Prior to joining White & Case, Mr. Mechanick was a senior executive for nearly twenty years with COMSAT Corporation and later Lockheed Martin Global Telecommunications, where he led U.S. efforts resulting in the privatisation of the International Telecommunications Satellite Organisation (INTELSAT) and the International Mobile Satellite Organisation (INMARSAT) and the creation of New Skies Satellites, NV, as an INTELSAT spin-off company. Mr. Mechanick served as the U.S. Governor to INTELSAT from 1995 to 2001 and as Chairman of the INTELSAT Board of Governors during the pivotal year leading up to that organisation's privatisation. Mr. Mechanick has previously served as the President and Chairman of the Society of Satellite Professionals International and has been elected to membership in the IISL. Mr. Mechanick received a BSFS degree (Foreign Service) from Georgetown University in 1973 and a JD degree from the University of Pennsylvania Law School in 1976.

*Christian Roisse* has been Executive Secretary of the European Telecommunication Satellite Organisation (EUTELSAT) IGO since 2005. Prior to this he was General Counsel of Eutelsat S.A. and before the organisation's restructuring he was Director of Legal Affairs at EUTELSAT for 13 years. With a Master's degree and postgraduate diploma in Public Law, he was previously Legal Adviser on business law, industrial law, telecommunications law, European and international law at France Telecom and at the French Department for Industry. He has made numerous presentations at international symposiums and written articles in the field of space activities, in particular on EUTELSAT and its restructuring process. He is a Founding Commissioner of the Broadband Commission for Digital Development established by the ITU and UNESCO, a member of the IISL and the European Centre for Space Law, and he serves on the Board of Directors of the Association for the Development of Space Law in France.

*David Sagar* was a Legal Consultant to the International Maritime Organisation (IMO) in London, 2008–2011. He was Senior Legal Attorney to the International Mobile Satellite Organisation (INMARSAT) from 1980–1999, and LLM Course Director in Satellite and Space Law at University College and Queen Mary college, London University from 1996–2006. He holds LLB and LLM degrees from the University of Sydney (NSW), and is a Solicitor of the Supreme Courts of New South Wales and England. Earlier work included the practice of commercial law in Sydney; legal adviser at the International Commission of Jurists (ICJ) in Geneva; Registrar of the

Commission of Inquiry into Racial Discrimination in the Public Services of British Guiana, 1965; and legal adviser at the Commonwealth Secretariat in London, 1970–1978. He is a member of the IISL, and a lecturer of summer courses at the European Centre of Space Law (ECSL) and other venues on various space law subjects including the activities and privatisation of the international satellite organisations, problems of space debris, the working of the ITU, contractual and procurement issues in space-related activities, and the use of space for peaceful purposes. He has contributed articles on space-related topics to the *Journal of Space Law* and other publications, and has been an adviser to the Netherlands Government on the adoption of national legislation for the implementation of the space treaties.

*Frans von der Dunk* is currently Harvey and Susan Perlman/Alumni Professor of Space Law at the University of Nebraska-Lincoln (UNL), as well as Director of the space law and policy consultancy Black Holes in the Netherlands. Prior to his appointment at UNL he has served on the Faculty of Law of Leiden University in various capacities in the field of public international law and space law for some 20 years. Prof. von der Dunk obtained his Doctorate in Space Law at Leiden University in 1998, and has written over 120 articles in the field of space law and policy. He has been awarded *inter alia* the IISL Distinguished Service Award in Vancouver in October 2004, and the IAA Social Science Award in Valencia, in October 2006. He is, amongst many other functions, ECSL Board Member, Member of the IISL Board of Directors and Honorary Professor at Gujarat National Law University, Gandhinagar, India. Prof. von der Dunk has advised many governments, international governmental and non-governmental organisations, space agencies and companies on issues of space law and policy.

# Index

www.ingramcontent.com/pod-product-compliance
Lightning Source LLC
LaVergne TN
LVHW102103300926
849161LV00001B/10
*9789004257061*